Through the Looking Glass

Telescopes

Through the Looking Glass

Marvin Bolt

The exhibition *Telescopes: Through the Looking Glass* opened at the Adler Planetarium & Astronomy Museum in 2009.

Section Dividers
Part I Larry A. Ciupik, Astronomer, Adler Planetarium
Part II © Akira Fujii/David Malin Images
Part III NASA, ESA, M. Robberto (Space Telescope Science Institute/ESA) and the Hubble Space Telescope Orion Treasury Project Team
Part IV NASA, ESA, T. Megeath (University of Toledo) and M. Robberto (STScI)

Webster Institute for the History of Astronomy
Adler Planetarium & Astronomy Museum
1300 South Lake Shore Drive
Chicago, Illinois 60605

Library of Congress Cataloging-in-Publication Data

Adler Planetarium & Astronomy Museum.
Telescopes: Through the looking glass/by Marvin Bolt.
p. cm.
Includes bibliographical references.
ISBN 1-891220-06-3
1. Adler Planetarium & Astronomy Museum – Exhibitions. 2. Telescopes – History – Exhibitions. 3. Cosmology – History – Sources – Exhibitions. 4. Astronomy – History – Sources – Exhibitions. 5. Telescopes – Illinois – Chicago – Exhibitions. I. Bolt, Marvin, 1963- II. Title.

QB88.A35 2009
522'.207477311 – dc22

2008053480

Acknowledgments

Developing this exhibition and catalogue involved the efforts and support of many team members.

The Adler Planetarium gratefully acknowledges Barry MacLean for his continued friendship and support of the Adler collections.

The Adler Planetarium is grateful to its Board of Trustees for supporting this exhibition. Special thanks to Nicholas J. Pritzker for his support of the modern telescopes and cosmology section of this exhibition.

The generous support of Marjorie and Roderick Webster strengthened the Adler collections to make this exhibition possible.

Support was also provided by the National Aeronautics and Space Administration (NASA).

The exhibition was enhanced by a few select loans: Gerald Fitzgerald provided us with a very rare copy of Galileo's provocative *Sidereus Nuncius*; Martin Roth, Director General of the Staatliche Kunstsammlungen Dresden, and André van der Goes, Director of the Kunstgewerbemuseum, Staatliche Kunstsammlungen, Dresden, Schloss Pillnitz graciously allowed us to borrow what may be one of the two oldest surviving telescopes in the world; and finally, Michael Korey, curator at the Mathematisch-Physikalischer Salon in Dresden, made the overseas loan possible. To each, I express my gratitude for accommodating my request and for enabling us to put together this grouping; as far as I know, no other exhibition has been graced with such a grouping of the world's earliest telescopes.

Several scholars, including Barb Becker, Paul Gehl, Barb Korbel, Ruth Norton, Rachel Freeman, Ed Kibblewhite, and others helped with the loans or provided expert knowledge that worked its way into one of more entries. Others have laid the foundation for building my understanding of the optical and material components of telescopes. Duane Jaecks and Gene Rudd have given me superb mentoring about testing telescope optics over the years. Peter Louwman and Rolf Willach provided me with access to their own extraordinary collections, giving me a much better understanding of early telescopes. Colleagues throughout Europe (Jim Bennett, Museum of the History of Science, Oxford; Liba Taub, Whipple Museum, Cambridge; Paolo Brenni, Giorgio Strano, and Paolo Galluzzi, Museum of the History of Science, Florence; Richard Dunn and Gloria Clifton, National Maritime Museum, Greenwich; Marinella Calisi, Monte Mario Museum, Rome; Michela Benvegnù, Lux Ottica, Agordo, Italy; Fabrizio Bònoli, Bologna, Italy; Christian Sichau, Deutsches Museum, Munich; Michael Korey; and others) as well as in China (Jin Zhu, Beijing Planetarium) have provided me with access to a wealth of telescopes that have informed my understanding of the items in this catalogue.

Our material culture consultants, including Tom Fuller, Tracy Kostenbader, Earl Lock, Chris Conniff-O'Shea, and Marty Sears, once again provided great skill to conserve and display our treasures. Steve Pitkin did superb photography, and Studio Blue creatively designed this catalogue.

This exhibition required diverse expertise from a talented Adler team: Christine Minerva, Michelle Nichols, Rosie Roche, Karen Carney, and Sue Wagner kept our educational goals in mind and our plans in focus while helping shape the exhibit and its programming; Mark Hammergren and Michael Smutko gave helpful advice about presenting the wonders of modern technology; Steve Woods, Mike McGowan, Rich George, Jerome

Lane, and Isaac Morales used their skills to make the physical exhibit come to life; Katie Peterson, Angelique Rickhoff, Nancy Ross, Julieta Aguilera, Doug Roberts, and Mark SubbaRao developed media pieces to animate the modern section of the exhibit and link it to the Space Visualization Lab; Mark Paternostro and Craig Stillwell provided artistic support; Christina Turner did heroic work identifying, locating, and organizing artifacts and images, as well as writing preliminary drafts of many entries; Jill Postma helped locate books and bibliographic data; Devon Pyle-Vowles and Jennifer Brand made sure our artifacts were conserved, photographed, and otherwise cared for, and provided great insights into the layouts of the exhibition; Shera Street and Beth Azuma gave excellent overall project management, while Misty DeMars and Jodi Lacy provided superb organizational control over numerous processes to keep the project, and me, on target; Bruce Stephenson edited this catalogue with great skill and insight; Ginevra Ranney, Gena Johnson, Ave Costa, and Charles Katzenmeyer provided wise counsel for raising financial support; Paul Knappenberger and Marge Marek gave great encouragement and helped identify funding opportunities; Molly O'Connell and Sarah Beck found innovative ways to spread the news about this project. Last, but not least, I take my hat off to Julian Jackson, Adler's director of experience design; it has been a great pleasure to work alongside him and to see creative ideas emerge and take shape.

To each supporter, scholar, and team member, I express my sincere gratitude and deep appreciation. I hope that you and each reader enjoy your trip *Through the Looking Glass* and find continued inspiration to explore our spectacular Universe.

Marvin Bolt, Ph.D.
Director, Roderick and Marjorie Webster Institute
for the History of Astronomy

Preface

From the simple lenses of the world's earliest telescopes 400 years ago to the complex computer-driven mirrors of current telescopes, these tools have gathered information about our nearest astronomical neighbors and the most distant objects in the Universe. Readers taking a trip *Through the Looking Glass* will enjoy a variety of these telescopes, travel to the worlds they enable us to see, and encounter the models that explain the structures of the cosmos.

The catalogue in your hands enriches the Adler's exhibition in celebration of the United Nations' declaration of 2009 as the International Year of Astronomy. This worldwide program encourages people around the globe to view the heavens with a telescope, to provide an inexpensive but good-quality telescope for anyone interested in obtaining one, and to educate the world about the wonders of the Universe that telescopes enable us to enjoy.

Unless otherwise noted, the artifacts featured in this exhibition and catalogue belong to the extraordinary collections of Chicago's Adler Planetarium & Astronomy Museum. Our selections highlight the evolution of telescope technology, the cosmological impact of this evolving technology, and some of its connections to popular culture. We hope that you enjoy your journey through the looking glasses of the past, present, and future.

Introduction

The setting is September 2008, on a train car en route to Middelburg, a town in the southwest corner of the Netherlands. I notice a young Dutch woman peering intently into her mirror, perceiving a few blemishes, and addressing them with cosmetics. At the moment, I am on my way to a conference about the telescope's invention 400 years ago in this town. It suddenly strikes me that we are, in fact, involved in similar projects.

She is using an optical device specially designed for observing parts of the nearby Universe, and supplementing that with tools that will impose more order on what can be observed. I am interested in an optical device designed for observing more distant parts of the Universe, a device supplemented with tools that impose some order on those observations. Her optical mirror is part of cosmetology, while telescope mirrors and lenses belong to cosmology. Each discipline shares the same two Greek words, *cosmos* (meaning "order") and *logos* (meaning "word"). Each resulting English word means the study of order; in the first case, the order concerns a face or a body, in the second the Universe itself.

My train companion is trying to make more order out of perceived chaos. She has a certain perception of what her face, her microcosm of the Universe, should look like. A few things that she sees in her mirror are anomalies, things that don't quite match her view of what the grand picture should be. So she has obtained a few instruments that help her to fix these problems, blending them carefully into her theoretical model so that these anomalies no longer stand out. She has at her dispersal a host of these specialized tools, and if she is missing something, she has access to a cosmetic counter, where she can find tools produced by an international fashion team who are willing and ready to help her. The finished result will be her local variation of a standard model of a face that has been developed or promoted by this international team. Even though that standard model changes – sometimes the fashion is to have eyes or some other feature highlighted in one way rather than another – the actual features of the face provide the background, content, and constraints of what a cosmetologist must work with.

My colleagues in cosmology likewise try to find the order underlying the apparent chaos and sometimes blurry details of the Universe. They have developed a model of the Universe to explain what they see, but sometimes they also notice anomalies, features that don't fit their model. So they build instruments, either by themselves or in cooperation with a national or international team, to explore those features. As they observe and understand these features better, they find a way to include them and blend them more seamlessly into their model, or come up with a new model. This model, an explanation of the elements, patterns, and behaviors of the Universe, shows cosmic order, and plays an important role in the discipline of cosmology. In the end, a cosmologist has to account for and work with the actual features of the Universe, as difficult, puzzling, or frustrating as they might be.

While only a few people make a living at cosmetology or cosmology, there is an important sense in which everyone participates in these activities. We all observe the world, try to make as much sense of it as we can, and develop a worldview that best explains the Universe as we see it. From time to time, we encounter people and events not fitting into our tidy schemes, and we have to adjust our vision and understanding of the world.

But not just anything goes. In cosmology and the natural sciences, evidence provides some measure of objectivity that counters mere opinions. In these and all academic disciplines, explanations are analyzed and examined according to agreed upon processes and are subject to peer review, providing an effective way to develop stories and theories that have warrant and merit.

In this exhibit, we concentrate on one tool that has transformed our observations of the world and the cosmologies or models that explain what we have seen through telescopes. In many instances, the construction materials and beautiful decorative elements of these telescopes further demonstrate the connection between cosmetology and cosmology. While engaging with the artifacts and images of the Adler's exhibition celebrating the International Year of Astronomy 2009, we hope that you will enjoy the scientific and broader cultural aspects of the evolution of the telescope and of our understanding of the Universe.

Historical background

Although lenses and mirrors of some sort were known even in antiquity, the first true eyeglasses or spectacles appeared around 1285 in Italy. More than 300 years elapsed before such spectacles yielded a working telescope. The reason for this delay has long puzzled investigators, but in his recent description of the path leading to the telescope's invention, Rolf Willach has proposed an intriguing solution. He suggests that lenses for eyeglasses could not be made with sufficient precision prior to around 1600, and that the key step for producing a telescope required a "mask" that covered up the less precisely shaped edges of lenses. Once this secret was revealed in 1608, the device spread quickly around Europe and soon into Asia. As Albert van Helden puts it, "in the summer of 1608, no one had a telescope; in the summer of 1609, everyone had one."

With this device, "everyone" was soon spying on neighbors and on military and economic competitors. Competition soon extended throughout the Universe. The Ptolemaic model of the cosmos was severely challenged, but the Copernican system did not immediately convince people; it was too counterintuitive. By the end of the 17th century, when the Sun-centered view of Copernicus was accepted, the telescope showed that the Universe is unimaginably large. As additional observations and tools revealed not only more stars but also the actual material composition of the most distant bodies in the Universe, careful investigations and interpretations suggested radically different models of the Universe and of its size and history.

Today's telescopes bear little resemblance to those small pieces of glass used by Galileo and his contemporaries. Those lenses have long given way to enormous mirrors and mirror arrays; the human eye has been superseded by much more sensitive detectors that feed complex computer systems with more data than any human could process in a lifetime. Our modern observatories house telescopes on mountains and on spacecraft orbiting the Earth and traveling throughout the Solar System, detecting not just optical wavelengths but energies throughout the electromagnetic spectrum, as well as subatomic particles.

And yet, the questions we ask today have not changed in principle from those posed in centuries past. We are still interested in finding and explaining the cosmic patterns we observe, and determining our place in the Universe. Telescopes have played a central role in this pursuit, and with this exhibition and catalogue, we hope you will enjoy your journey through the looking glasses of the past, present, and future.

Exhibition themes

The exhibition and catalogue address four themed zones. The first, the pretelescopic zone, addresses ways in which people have looked at the sky and tried to make sense of it, using their surrounding landscapes or relatively simple tools to develop an understanding or model of the Universe. Zone two presents the invention of the telescope, the challenges it brought to the Earth-centered Universe, and the beautiful craftsmanship and ornamentation of some of the earliest surviving examples in the world. In zone three, the technical challenges of improving telescopes led to variations in design and materials; telescopes also became popular devices with brand-name recognition. Zone four displays the culmination of the refracting telescope and the emergence of spectroscopy, leading to the marvels of modern telescopes: some see wavelengths beyond the optical realm, others detect invisible particles, a few compensate for atmospheric turbulence, while still others travel beyond the Earth's atmosphere into space.

Adler collections

Telescopes: Through the Looking Glass makes extensive use of the Adler's collections of historic scientific instruments, rare books, and works on paper. Because museum founder Max Adler wanted to replicate in Chicago the opportunities afforded by the Deutsches Museum in Munich, he purchased about 500 scientific instruments from an antiques dealer in Amsterdam in 1929. Over the course of eight decades, the Adler collections increased more than fourfold. Roderick and Marjorie Webster provided extraordinary curatorial service to raise the extent, quality, and profile of these collections. Today, the Adler's Webster Institute for the History of Astronomy provides a home to artifacts ranging from the 12th through the 21st centuries. Readers can find more information on the Adler's collections at http://www.adlerplanetarium.org.

A note on the dates of artifacts

The dating of historic artifacts is rarely straightforward and often involves a lot of skilled detective work. Sometimes the clues are pretty obvious, such as when the date is stamped right on the object. Quite often, though, one has to make an informed estimate based on a wide variety of data, such as the name or signature of the maker or author, or the aesthetic style of the shape, paper, leather, or decorative elements on the object. This often requires comparison with similar objects in different collections or museums, with illustrations in a contemporary book, or with documentation that provides evidence.

In this catalogue, we provide our best estimates for the dates of these objects. With a known date, it will be listed without a modifier ("1744"), but when that is not possible, we will provide a date usually to within a decade or quarter-century. Such a date listed will have "c." (Latin for *circa*, meaning "approximately") in front of it. As we continue to learn more about these objects and their dates, we will update our on-line collections database.

The Era of the Naked Eye

It seems natural to see the Earth as the center of the Universe, and to believe that everything revolves around it.

Every culture in human history has observed the events playing out on the celestial stage and has also tried to make some sense of them. By using landscape features or specially made buildings or markers to watch the appearances, disappearances, and seasonal locations of the Sun and the Moon, as well as of the planets and stars (which were not always explicitly distinguished from each other), people have observed the patterns of the heavens and often interpreted them as reflecting the actions of divine beings. The resulting stories explain the creation of the world and the events taking place in it.

In cultures that derive from the Babylonian, Greek, and Roman traditions, astronomy has a tradition of using sophisticated mathematics to describe those patterns. This tradition became a part of the Islamic world shortly after the time of Mohammed and was reintroduced into Europe prior to the 12th century. Precisely marked astronomical tools, such as the astrolabe, which enabled users to find the time of day or night and to determine the times for prayer, also entered Europe at this time.

So did the armillary sphere, already used in China and in the Islamic world, though the earliest known surviving image dates from ca. 1400 (see No. 2). The armillary sphere portrays the motions of the heavens as seen from Earth. The sky as a whole appears to rotate once per day. Whereas the stars remain fixed with respect to each other, the Sun, Moon, and planets all seem to wander against the starry background. It seems natural to see the Earth as the center of the Universe, and to believe that everything revolves around it. This geocentric or Ptolemaic model provided the standard view of the Universe from the time of the Greeks (from about 350 B.C.) until about 1700.

The broad history of the telescope begins in the late medieval period. Within 150 years after the introduction of eyeglasses to correct vision, they became available even for ordinary citizens in spectacle-making centers such as Florence and Venice in Italy and Nuremberg and Regensburg in Germany. In particular, the development of the printing press in the 1450s led to the widespread popularity of eyeglasses. But because their lenses could not be shaped consistently over their entire surfaces, eyeglasses could not be combined together to create a clear, sharp image larger than one obtained with just the unaided eye.

As a result, until 1608, astronomy involved the same processes it had during the previous centuries and millennia to observe the same events in the sky. It was the job of the astronomer to track the motions of the Sun, Moon, planets, and stars. The astronomer made accurately marked tools to observe, using only the unaided eye, the positions of the heavenly bodies, to create geometric models that explained those positions, and to teach the motions of the heavens to students.

Gaius Julius Hyginus

De mundi et sphaerae ac utriusque partium declaratione

Venice, Italy, 1517 PA6445 .H8A3 1517

Gaius Julius Hyginus (64 B.C.–A.D. 17), likely a native of Spain, was a librarian and prolific author, though most of his works have been lost. Only two mythological treatises survive in connection with his name, though many scholars believe they are the work of a writer (more than a century later) who abridged the works of Hyginus. One of these treatises, *Poeticon astronomicon*, recounts ancient myths about constellations. These words survived only in manuscript copies until Erhard Ratdolt first printed a copy in 1482; several reprints and editions appeared in the following decades.

De mundi et sphaerae en declaratione is part of this text, which begins with a discussion of some of the technical terms used by astronomers. In this frontispiece, the armillary sphere and astrolabe represent the tools of the astronomer and convey the authority and knowledge granted to the ruler seated on the throne, with the zodiacal constellations and stars looking on in approval.

Bartholomaeus Anglicus

Manuscript fragment from *Propriétés des choses*

Paris, France, c. 1400 P-270

The armillary sphere, known from Greek antiquity, is sometimes credited as having been invented by Eratosthenes, a librarian from Alexandria perhaps known best for his estimate of the Earth's size in the 3rd century B.C. This device was also known in ancient China, though most likely from a few centuries later. It was used as a teaching device in Greece and as an observing instrument elsewhere; this was its function in China and also in the medieval Islamic world beginning in the eighth century. It came to Europe around 1000 as a result of efforts by the future Pope Sylvester II. It symbolizes the Earth-centered Universe, and it still serves as a very useful tool for understanding the heavenly motions that are visible with the naked eye.

English Franciscan priest Bartholomew (c. 1203–1274), who was connected to the universities in Paris and Oxford, authored an important work *On the Properties of Things*, which addressed theology, physiology, medicine, physics, and much more in 19 books. It served as the first medieval encyclopedia and as a model of future encyclopedic efforts. Its enormous popularity can be recognized by its appearance in numerous manuscripts and translations throughout the medieval era; it also appeared in more than a dozen printed versions prior to 1500, including ones in English, French, and Spanish.

This manuscript fragment of Bartholomew's encyclopedia features one of the earliest known illustrations of an armillary sphere, the iconic model of the medieval geocentric Universe. Along with the astrolabe, the armillary sphere was frequently included in portraits until the time of the Renaissance to convey wisdom, knowledge, and learning.

Christoph Scheiner *Disquisitiones mathematicae* **Ingolstadt, Germany, 1614 QB41. S32 1614**

The *Disquisitiones* is, officially, a dissertation defended by Johann Locher, a student of Scheiner. Most likely, though, according to the custom of the time, it was actually written by Scheiner. It contains numerous important illustrations, including some of the earliest ones of a telescope and a very early map of the Moon, the first to give any significant topographic details. Despite his support for Galileo's observations of Venus and Jupiter, he disagreed vehemently on their cosmological significance.

The *Disquisitiones* provides illustrations of several cosmological models, along with a commentary on their relative merits and difficulties. Scheiner always maintained his opposition to the Copernican heliocentric theory and to any notion of an infinite Universe, represented in his volume by a full-page woodcut of "*Chaos infinitum ex atomis*" surrounding the sphere of fixed stars, itself surrounding the planetary system with Earth at its center. Over the next several pages follow discussions of several other cosmologies: the unacceptable Copernican account, nicely illustrated with the Sun and Earth highlighted; a geocentric version by Italian Girolamo Fracastorius (1478–1553); yet another geocentric model, this one by fellow Jesuit Christopher Clavius (1538–1612); and finally, the hybrid version of Tycho Brahe (1546–1601), which Scheiner defended. Other illustrations feature the Moon, sunspots, Jupiter's moons, Saturn, and phases of Venus. The latter illustrations also provide some of the earliest details of contemporary telescopes.

SYSTEMA CLAVIANVM.

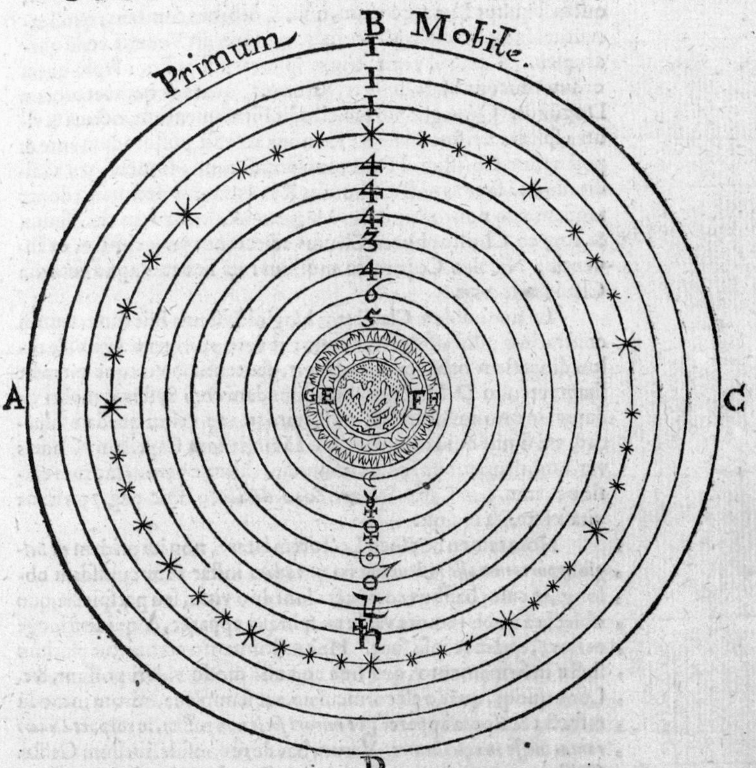

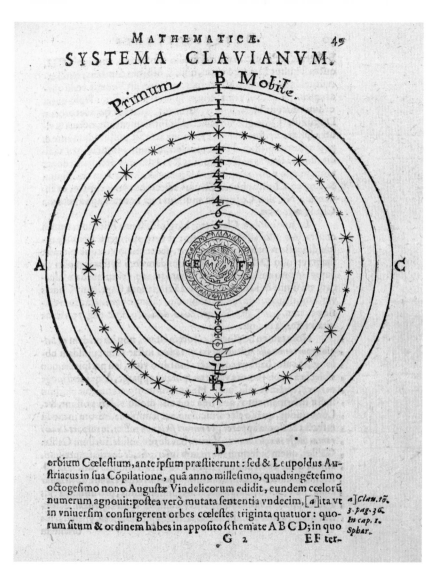

orbium Cœlestium, ante ipsum præstiterunt: sed & Leupoldus Austriacus in sua Cõpilatione, quã anno millesimo, quadringẽtesimo octogesimo nono Augustæ Vindelicorum edidit, eundem cœlorũ numerum agnouit: postea verò mutata sententia vndecim, [a] ita vt in vniuersim consurgerent orbes cœlestes triginta quatuor: quorum situm & ordinem habes in apposito schemate A B C D; in quo

a] Clau. tõ. 3. pag. 36. In cap. 1. Sphær.

G 2 E F ter-

John Peckham *Perspectiva Communis* **Nuremberg, Germany, 1542**

John Peckham (c. 1230–1292), a great advocate of conducting experiments, was part of a small group of 13th century medieval scholars who formed the perspectivist tradition in optics. They were interested in understanding how vision takes place and how images formed in the eye and the brain, including how this took place with the use of mirrors or glass spheres. Peckham, like most scholars of this time, assumed that visual rays were emitted by the eye and then returned to the eye after encountering an object. This woodcut of several centuries later illustrates Peckham's explanation of the operation of the eye. Peckham, the Archbishop of Canterbury at the end of the 13th century, opposed some of the views of St. Thomas Aquinas and spent much of his time trying to reform church politics.

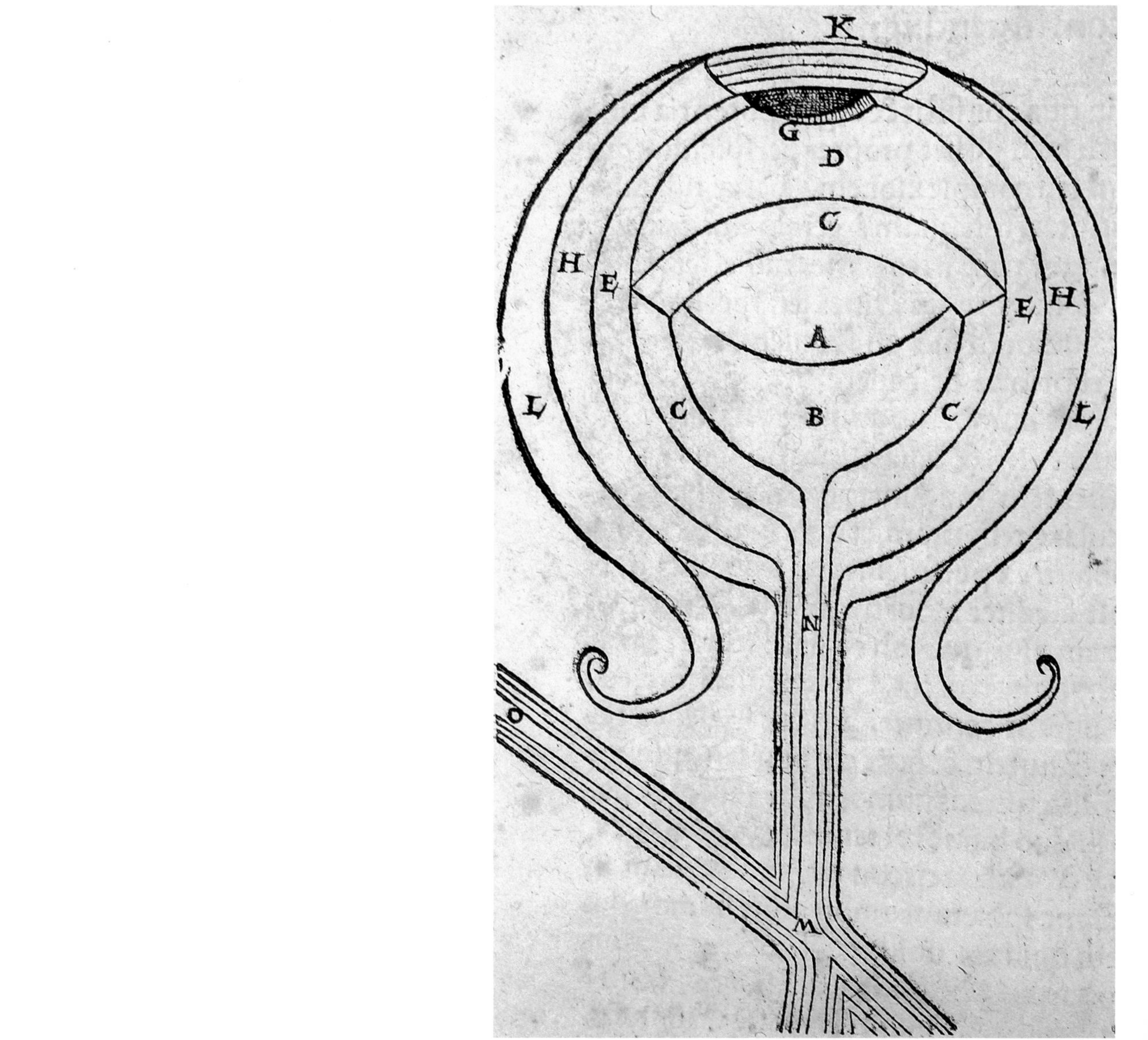
K
G
D
C
H
E
E
H
A
L
C
B
C
L
N
O
M

Benjamin Tanner **"The Observatory at Peking" from *Payne's Geography*** **New York City, United States, c. 1800 P-51**

Astronomical records in China go back at least 2,500 years. One of the world's oldest surviving observatories, the Beijing Ancient Observatory, stands atop a 50-foot-high, 125 x 125-foot platform remaining from the ancient Ming Dynasty city wall. The current observatory site, dating to 1422, provided astronomers with tools to carry out astrological and calendaric duties to the Ming and Qing emperors.

After Jesuit astronomers demonstrated their ability to predict eclipses in the early decades of the 1600s, the Chinese emperor requested Ferdinand Verbiest to design and build a suite of six astronomical devices, though none of them were telescopic. In 1673, he carried out this task; in 1715 and 1744, other instruments were added. These devices now represent the limits of the astronomical measurements and activities of the pretelescope era. Around 1900, several of these instruments were brought to Europe, but were returned to China near the end of World War I. Today these large, sculptural instruments stand in marked contrast to the modernity of nearby buildings.

Benjamin Tanner (1775–1848), an American engraver, assisted with the publication of maps and banknotes in addition to engraving portraits of famous Americans and events in American history. This illustration appears in a richly illustrated volume of world geography.

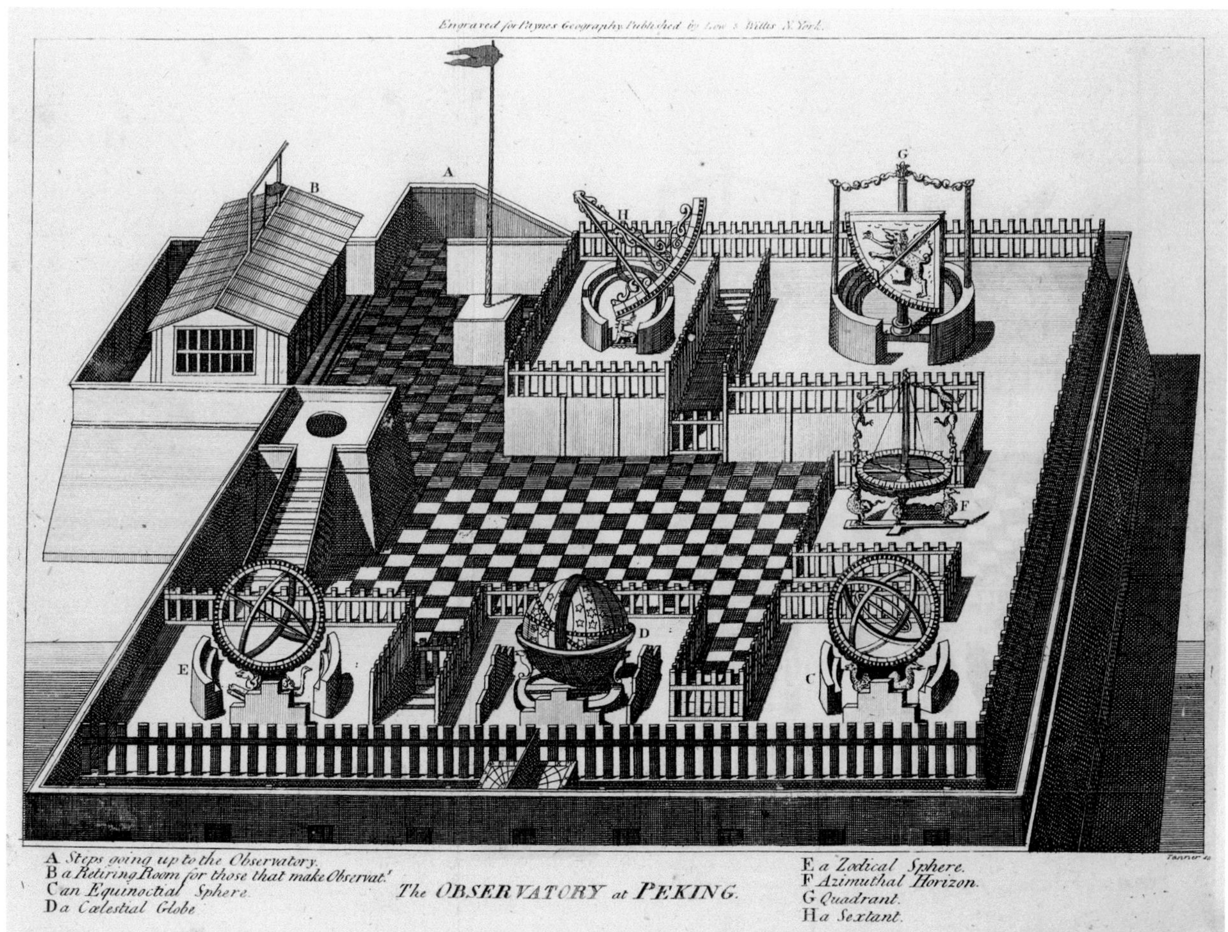
Engraved for Paynes Geography Published by Low & Willis N. York.
A
B
G
H
F
E
D
C
Tanner sc
A Steps going up to the Observatory.
B a Retiring Room for those that make Observat.s
C an Equinoctial Sphere.
D a Cælestial Globe
The OBSERVATORY at PEKING.
E a Zodical Sphere.
F Azimuthal Horizon.
G Quadrant.
H a Sextant.

Joan[nes] Blaeu

"Stellaeburgum sive Observatorium Subterraneum, a Tychone Brahe ..." from *Atlas Major*

Amsterdam, the Netherlands, 1662 P-125a

Joan[nes] Blaeu (1596–1673), a Dutch cartographer, was the son of Willem Blaeu, who had spent two years studying under Tycho Brahe (1546–1601). Tycho, a Danish nobleman, received an estate on the Danish island of Hven and funding to build an observatory, Uraniborg ("heavenly castle"), in 1576. In 1581, Tycho founded Stjerneborg or Stellaburg ("starry castle" in Danish and Latin, respectively), an underground version of Uraniborg.

Uraniborg was unsuitable for the careful measurements Tycho wanted to make. Instruments were protected from the bitter winds on the new site. Stjerneborg consisted of five chambers, each sheltering a large astronomical instrument and covered with a roof that could open. The largest, with a rotating dome, housed the great equatorial armillary sphere, while two others each contained two quadrants; the remaining two held an armillary sphere and a sextant.

After Tycho lost his funding from King Christian IV in 1597, he abandoned Hven, where he had also built a paper mill and printing press. The local population, never fond of Tycho and his autocratic rule of the island, destroyed both observatories soon afterward, looting the site for building materials.

Tycho's observatories were the last European ones not to include a telescope, invented a decade after his departure. Tycho did not accept the Copernican Sun-centered system or the Ptolemaic Earth-centered one, but proposed instead a hybrid system in which the planets orbit the Sun, which orbits the Earth. He made years of observations of Mars on Hven to determine the best cosmological model, but it was his student Johannes Kepler who used Tycho's data to formulate his laws of planetary motion and to confirm the Copernican system.

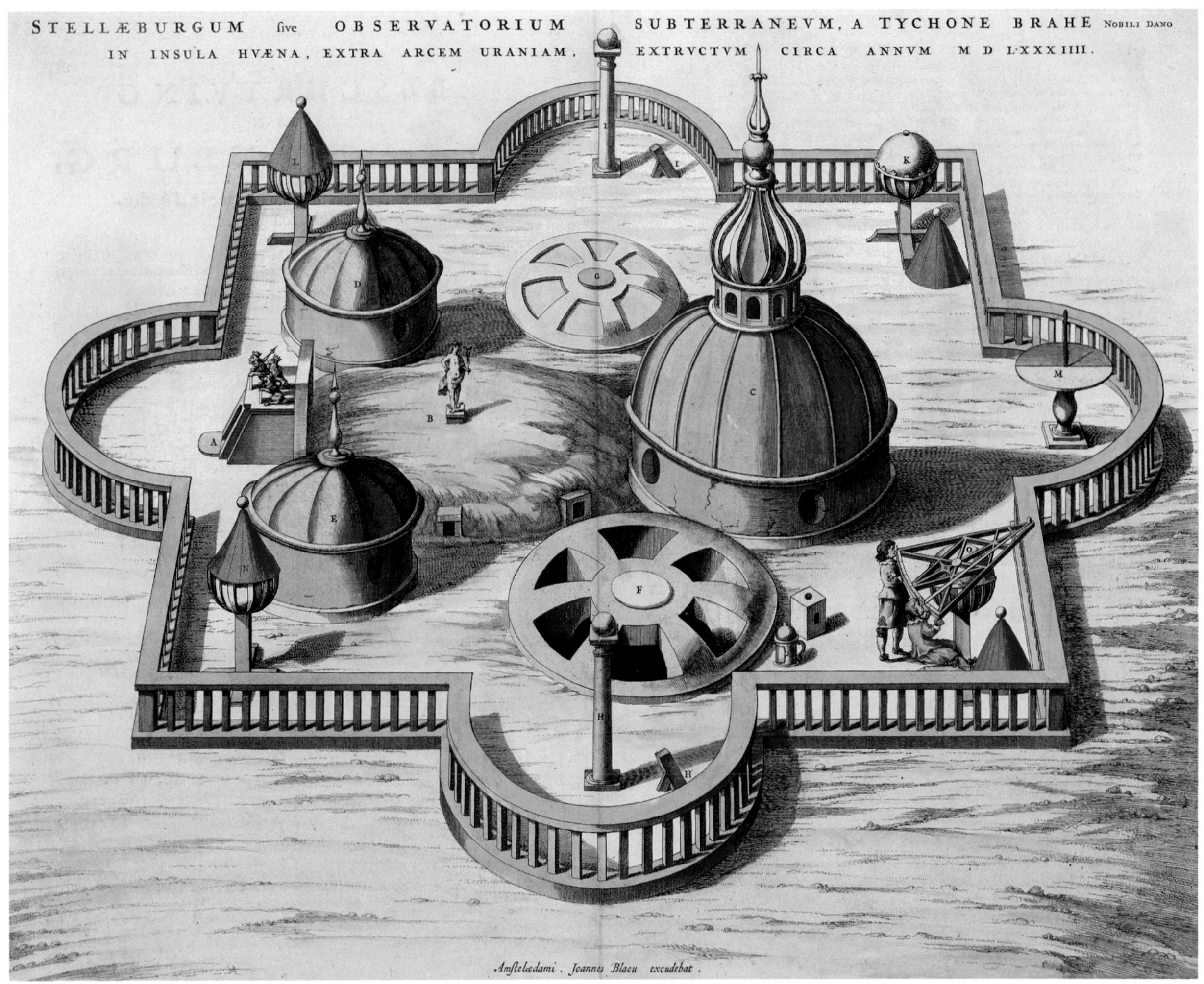
STELLÆBURGUM sive OBSERVATORIUM SUBTERRANEVM, A TYCHONE BRAHE NOBILI DANO
IN INSULA HVÆNA, EXTRA ARCEM URANIAM, EXTRVCTVM CIRCA ANNVM M D LXXXIIII.
Amsteleedami. Joannes Blaeu excudebat.

Galileo and Early Telescopes

Even the stars that ordinarily are invisible to our sight ... can be seen by means of this instrument.

The first documented demonstration of the telescope took place in The Hague on September 25, 1608, when German spectacle-maker Hans Lipperhey presented it to Dutch Prince Maurits, his younger brother Prince Frederik Hendrik, and the Spanish commander, Marquis Ambrogio Spinola. That month, a peace conference in The Hague included officials from the Netherlands and Spain, as well as several other countries, including Siam (present-day Thailand). Shortly following the completion of the conference, the Asian visitors left, and a small newsletter appeared in French describing the visit of the ambassadors from Siam.

After a description of various diplomatic details, there follows the first printed description of the telescope (though it did not use that word, which first appeared more than two years later at a banquet in honor of Galileo). The report mentioned that with the use of this device, one could see far things as if they were nearby; in particular it mentions that a clock about 12 miles away (in Delft) and church windows about 25 miles away (in Leiden) could be viewed clearly. Spinola quickly noted with some alarm the military use of this device. He informed Prince Hendrik that his own Spanish forces would now be in danger because they could be seen from a great distance; Hendrik calmly replied, "We will forbid our men to shoot at you."

Perhaps even more significantly, this newsletter points out that "even the stars that ordinarily are invisible to our sight ... can be seen by means of this instrument." In short, the astronomical significance of the telescope was noted from the outset. News of this device emerged in October in The Hague and in Paris, and the following month in Lyon, France. The telescope as a toy appeared in October or November 1608 at the Frankfurt book fair and was in spectacle shops in Paris the following spring; the telescope reached Japan by September 1613. Recent research has shown that at least six people had been observing the heavens intentionally before Galileo, who heard of the telescope many months later and built a successful telescope for the first time in July 1609.

It is possible that before 1608 several people put together lenses that enabled them to see distant objects as if they appeared closer. Indeed, Roger Bacon, Thomas Digges, William Bourne, Saccharias Janssen, and others all made such claims or have had claims made for them. In each case but the last, the claims do not hold up for their having made even one useful device; in Janssen's case, he did not make his first one until a decade or more after Lipperhey did.

To be credited with inventing such a device, it is necessary to be able to understand the invention sufficiently that it is possible to make additional examples of it. For this reason, Lipperhey is by far the most solid contender for being honored as the inventor of the telescope. He lived in Middelburg, the Netherlands, where the Dutch cosmetologist and I were heading 400 years later, and where the story of this catalogue begins. Not only did Lipperhey provide a working instrument in September, but he also made three additional telescopes – in fact, binocular telescopes consisting of two parallel single telescopes bound together. It is extremely difficult to make binocular telescopes; his success led to a reward of 300 guilders for the first instrument and 900 for the three binoculars. It would be quite some time before others were able to produce adequate binoculars and nearly three centuries before high-quality examples would be produced on a commercial scale.

Galileo

Istoria e Dimostrazioni Intorno Alle Macchie Solari e Loro Accidenti

Rome, Italy, 1613 QB525 .G13 1613

Recent scholarship suggests that Englishman Thomas Harriot was the first to observe sunspots, followed closely by Galileo, at the end of 1610. Others soon followed suit: Dutch observers Johannes and David Fabricius and German astronomer Christoph Scheiner viewed them in March 1611. Johannes Fabricius beat everyone to the press with his *De Maculis in Sole Observatis* in 1611. Scheiner's *Tres Epistolae de Maculis Solaribus Scriptae ad Marcum Welserum* appeared in January 1612, with Galileo soon receiving a copy.

In the summer of 1612, Galileo decided to turn his telescope to the Sun for systematic observations. He had earlier noted the "imperfections" of sunspots. Now he charted their changing locations on the Sun and published his observations. He and Scheiner continued to exchange letters, but eventually their courteous exchanges led to disputes over the priority of their discoveries and their differing interpretations.

The frontispiece is remarkable for its depiction of two instruments improved and made well known by Galileo. On the left side, a *putto* holds Galileo's analogue calculator, the geometric and military compass; on the right, another *putto* appears to hold a trumpet that is, in fact, the earliest printed illustration of a telescope. This provides a tantalizing suggestion about the shape of early telescopes. The accuracy of Galileo's likeness can be seen by comparing it with the portrait done by the Medicean court painter, Justus Sustermans (on the opposite page).

Justus Sustermans

Portrait of Galileo

Unknown Location, 1635 P-214

Perhaps the most widely recognized portrait of astronomer Galileo Galilei (1564–1642) is this one, painted by Justus (or Giusto) Sustermans (1597–1681). Born in Antwerp, Belgium, Sustermans was a famous painter and portraitist in the service of the Medici court in Tuscany. His broad style used elements from the Baroque era of Flemish, French, Spanish, and Venetian painters.

Galileo's contributions to astronomy include the telescopic discoveries of the phases of Venus and the four largest satellites of Jupiter, and the observation and analysis of sunspots. Each of these provided support for the Copernican Sun-centered model of the heavens. Modern scientific endeavors named after him include the Galileo spacecraft, the first probe to orbit Jupiter. In his *Sidereus Nuncius*, which appeared in March 1610, Galileo suggested naming the new moons "the Medicean stars" in honor of the powerful Medici family. Just three months after showing these "stars" to Cosimo II de' Medici, Grand Duke of Tuscany, in April 1610, Galileo was appointed as "Chief Mathematician of the University of Pisa, and Philosopher and Mathematician to the Grand Duke" for life.

Galileo attracted much attention and drew considerable criticism from the Catholic Church in Rome for his defense of the Copernican system. In this portrait, likely done toward the end of his life, Galileo shows some evidence of the stresses relating to that defense.

Istoria e Dimostrazioni Intorno Alle Macchie Solari e Loro Accidenti
Rome, Italy, 1613
QB525 .G13 1613

Portrait of Galileo,
Unknown Location, 1635
P-214

Petro Borel

De Vero Telescopii Inventore

The Hague, the Netherlands, 1656
QB88 .B67 1656

Pierre (Petro) Borel's account of the invention of the telescope (and of the microscope) contains portraits of the two main claimants as the telescope's inventor, Hans Lipperhey (1570–1619) and Saccharias Janssen (c. 1580–c. 1638), each of whom was a spectacle maker in the Dutch city of Middelburg. Borel (c. 1620–1671), writing several decades after the actual event, sides with Janssen in this history of the first 50 years of the telescope, complete with important descriptions of how to grind lenses and make a telescope. Borel was misled, perhaps by nationalistic pride, to prefer the Dutch Janssen over the German Lipperhey. Recent scholarship shows that Lipperhey preceded Janssen, who was later convicted of forging coins. In his detailed examination of the telescope, its construction, and its use, Borel also announces Christiaan Huygens's proposed solution to the puzzling appearances of Saturn prior to the publication of Huygens's *Systema Saturnium* (see No. 19).

Berckman
J. v. Meurs sculp

Christoph Scheiner *Rosa Ursina* **Bracciano, Italy, 1629 QB521 .S3 1629**

Christoph Scheiner (c. 1573–1650), a Jesuit priest and astronomer who lived in Ingolstadt, Germany, first achieved fame in 1603 with his invention of the pantograph, a device for copying diagrams. He also made great progress in understanding the process of vision and showed the importance of the retina as the seat of vision and the optic nerve as the transmitter of images to the brain. After 1610, he lectured on many topics in mathematics and astronomy. In 1611, he began to use the telescope for his solar observations, which continued for at least two decades. He soon noted the presence of sunspots and published several studies that included numerous observations of them. This launched a long and bitter dispute with Galileo over the priority of discovering sunspots and over their interpretation, with Galileo arguing that these were imperfections on the Sun itself and Scheiner initially arguing that these were astronomical bodies passing in front of the Sun.

In *Rosa Ursina*, Scheiner provided detailed illustrations of sunspots and of his setup for how he observed them (bottom illustration). His technique for tracking the changing shapes and locations of sunspots set the standard for the next century. By now, Scheiner no longer defended the satellite sunspot thesis or the view of celestial perfection, but he never accepted Copernicanism, opting instead for the Tychonic model of the planetary system. In the top illustration, we find important details of early telescopes and how they were actually used.

The intricate details portrayed throughout the volume, such as on this copper plate engraving, required four years to engrave and print. Such an expensive process required patronage, in this case, provided by Count Orsini. His family emblem, the rose, appears in the four corners of the upper illustration; together, the rose and the bear (*ursa*) provide the title of this beautiful volume.

150.
IMPROBVM AVSVM
VERVM ITA SVBTILI ARGVMENTATIONE COMPREHENSVM
VT CREDERE PVDEAT NON
Teliofcopio
Immiſſione naturali.
Transmiſſione Refractoria ſimplici.
Helioſcopio
Reflexione.
Immiſſione Refractoria compoſita.
Maculæ et Faculæ ex uariis obſeruandi modis ſtabiliuntur.

Simon Marius

Mundus Jovialis

Nuremberg, Germany, 1614 QB661 .M37 1614

As did many others, astronomer Simon Marius (1573–1624), born near Nuremberg, Germany, in the town of Gunzenhausen (mentioned at the top of this illustration), entered into a dispute with Galileo. In this text, Marius describes the world of Jupiter and its satellites and claims to have seen them prior to Galileo having done so. It now seems most likely that Marius first observed them after Galileo, though independently. Ironically, the four moons named by Galileo as the Medicean stars are often known collectively as the Galilean satellites, but individually their "official" names are those given by Marius: Io, Europa, Ganymede, Callisto (*Mundus Jovialis*, 78). In this frontispiece, Marius is shown with a chemical apparatus in his left hand, a drawing instrument in his right, and a book and telescope on the table in front of him. Although the term "telescope" had been introduced in 1611, Marius continues to use the word *perspicillum*, seen at the bottom of this portrait.

SIMON MARIVS GVNTZENH. MATHEMATICVS ET MEDICVS ANNO M. DC. XIV. ÆTATIS XLII.

JNVENTUM PROPRIUM EST: MUNDUS IOVIALIS, ET ORBIS TERRÆ SECRETUM NOBILE, DANTE DEO.

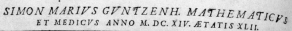

PRIMA PARS DE AMPLITUDINE MUNDI JOVIALIS, CONSIDERATIO UNIVERSALIS.

DEſcripturus hiſtoriam Mundi Iovialis, haud inconſultum duxi, totam libelli ſeriem in tres ſubdividere partes. In prima tractabitur univerſalis conſideratio hujus Mundi Iovialis, videlicet amplitudo ejuſdem, & quatuor in eo contentorum corporum magnitudo, & motus velocitas circa Iovem probabiliter determinabitur. In ſecunda particulares motuum differentiæ explicabuntur. In tertia omnia illa phænomena convenienti Theoria explicabuntur, quibus tandem tabularum compoſitio & uſus ſubjungetur, qui eſt principalis ſcopus totius hujus libelli. Ordiar itaq; ab univerſali conſideratione Mundi hujus Iovialis, à prima machinæ mundanę conditione omnib. mortalibus incogniti. Per diligentem poſſibilem, eamque diurnam obſervationem deprehendi Iovem continere in diametro propria 35. ſexageſimas quaſi, diametri terreſtris. Nam ſua diametro in media à terris

A diſtan-

Joseph Walker

"The Sun according to Fa[ther] Kircher"
Astronomy's Advancement

London, England, 1684 QB88 .A88 1684

Little has been written about Joseph Walker or his work, though this text is thought to be a translation of an anonymous French work. About Athanasius Kircher (1602–1680), though, far more is known. A Jesuit priest, Kircher contributed to many areas of study, including astronomy, geology, medicine, and, perhaps most important of all, Egyptology, of which he is considered the founder. He was mechanically inclined and designed several machines, including a magnetic clock, diverse musical instruments, and automata (robotic machines). He also wrote prolifically on equally diverse topics. He was especially interested in images and the projection of images for public display. Though not the inventor of the magic lantern, the forerunner of the slide projector, he improved it and made use of it. His interest in geology led to a study of fossils, earthquakes, and volcanoes, and he suggested that the Earth's internal fires are the cause of many geological events. In keeping with this view, Kircher proposed that sunspots are the results of solar volcanic activity, as shown in this beautiful illustration. This reasoning demonstrates one of the consequences of the Copernican worldview: If the Earth is simply one planet of many orbiting the Sun, then other planets may be similar to Earth in their topography, biology, and climate. Moreover, maybe even the Sun displays some of the same physical properties as the Earth.

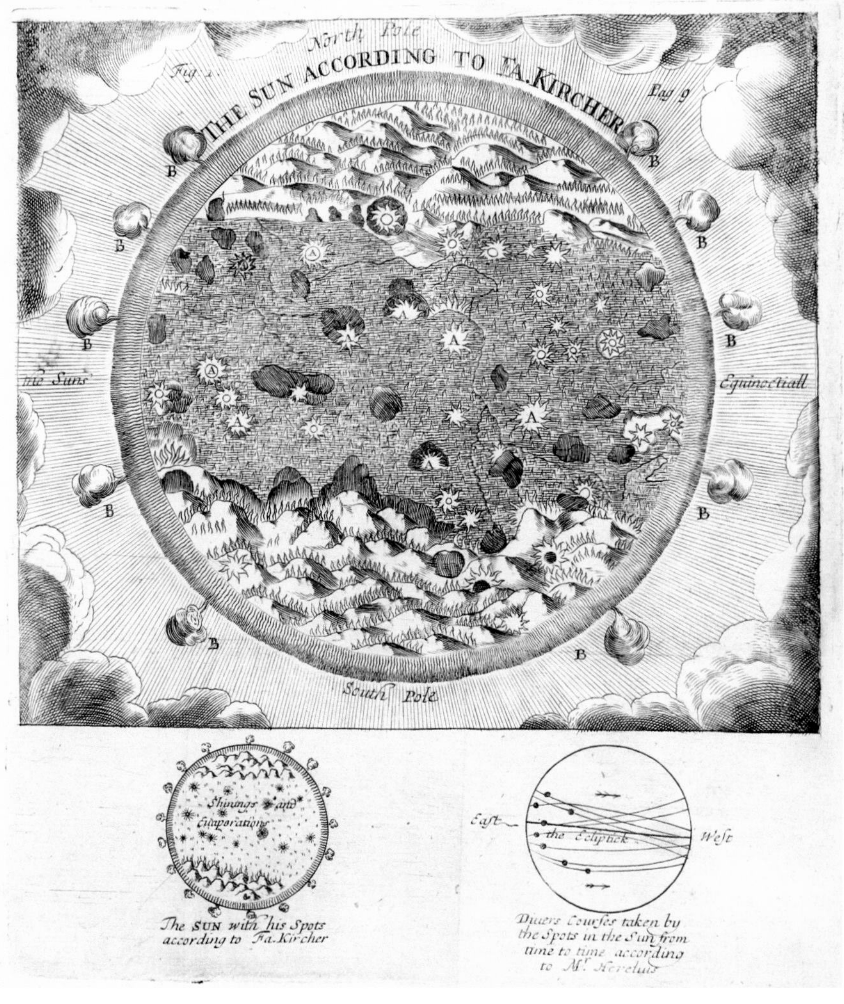
North Pole
Fig. 1.
THE SUN ACCORDING TO FA. KIRCHER
Pag 9
the Suns
Equinoctiall
South Pole
Shinings and Evaporations
East
the Ecliptick
West
The SUN with his Spots according to Fa. Kircher
Divers Courses taken by the Spots in the Sun from time to time according to Mr. Hevelius

Francesco Bianchini

Hesperi et phosphori nova phenomena

Rome, Italy, 1728 QB621 .B5 1728

Italian astronomer Francesco Bianchini (1662–1729) is best known for his work on reforming the calendar, especially the vexing problem of determining the correct date to celebrate Easter during any given year. His elegant solution involved building a meridian line in the Church of Santa Maria degli Angeli in Rome. His volume on Venus (named as Hesperus, the evening star, and Phosphorus, the morning star) contains beautiful illustrations of the telescopic views of Venus and its phases. Whereas Galileo and others saw these phases as proving the Copernican system, Bianchini slyly portrayed his planetary system with an unlabelled center.

Planetæ Veneris Phases conspectæ, et Maculæ ad instar Marium Lunarium detectę, necnon vertigo circa Axem proprium spatio dierum 24. Romę et Albani anno 1726 per Telescopia I. Campani Palmorum Rom. 88 et 94.
die 9 Februarii
die 14 Febr.
TAB. I
die 16 Febr
die 18 Febr.
Omnia exhibentur situ inverso, ut in telescopio apparebant.
Detexit, observavit, edidit
Franciscus Blanchinus Veronensis SS. D. N. Papę Prelatus Domesticus

Hieronymi Sirturi

Telescopium; sive ars perficiendi novum

Frankfurt, Germany, 1618 QB88 .S54 1618

Girolamo Sirtori of Milan provided the first extended discussion of telescope making, complete with a wonderful series of images showing the tools and techniques of how one could produce lenses and build a telescope. Early on in this volume, Sirtori notes that a telescope had reached a Count de Fuentes in Milan, Italy, in May 1609. This is when Galileo first heard of this device and a month before he made his own telescope.

Sirtori describes how to use a broken piece of a Venetian mirror as the starting point for making a telescope lens. Such mirrors at that time were ground and polished on a large iron plate; this already-polished side was left alone by the optician, who had only to grind and polish the other side to produce a lens of sufficiently high quality for use as a telescope lens. This illustration is the first schematic drawing of the optical properties of the telescope.

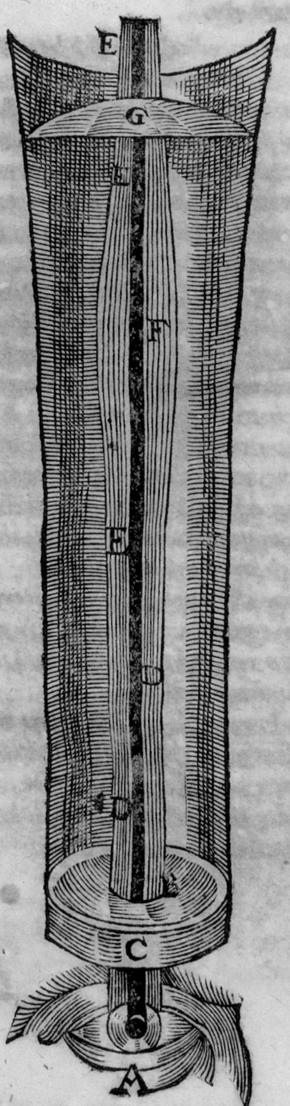

Tubus & obiter de refractione.

A. Pupilla oculi interior, ex qua prodit maſſa radiorũ vehementiſſima (Optici axem, & perpendicularem vocant) quæ irrefracta vtrumq; diaphanum peruadit, debilior tamen eſt quanto magis recedit ab oculo, ſed iuuatur forti viſiua quæ intendit aliud ſibi magis connaturale diaphanum pertranſire, & videre, vnde languerent antequam peruenirent ad lentem, niſi intenſa & dolorosa vi ſuſtinerentur. Vis igitur quæ procedit à cauo protẽdit quidem radios ad lentẽ conuexã, ſed vbi propius accedit, mutat naturam & vim ex natura conuexi acquirit, & quodãmodo trahuntur radij alioqui ſuapte natura defecturi.

Breuis viſus oculis myopes Græci, nõ eſt connaturale ideo non vident vt ſani.

Interior igitur pupilla A. ea eſt quæ radios tranſmittit irrefractos F. Reliqua pars oculi alios radios D. qui cauum ſpicillum C. egreſſi per foramen B. diuergũt, ſed ſuſtinentur natura irrefractorum radiorum, & vbi perueniunt in F. prope diaphanũ ſibi connaturale vi oculi, eaq; adiuta

Unidentified Maker **Refracting Telescope** **Italy, c. 1630 M-421**

Although it seems likely that thousands of telescopes were made prior to 1650, we know of approximately one dozen that still survive. This beautiful Italian example is the only survivor known outside of Europe and is thus the earliest in the Adler's collections. Dating from roughly 1630, this refractor also has an unusual form: Its main tube tapers towards the eyepiece end. Only seven known examples feature this "trumpet" shape.

The main tube is constructed from pasteboard, that is, glued layers of paper, covered with tan leather that has been gold-tooled with small designs of birds, flames, flowers, and other motifs. The telescope's aperture disk and five draw-tube ferrules are also covered with similar tooled leather that was likely more reddish in color, as is still observable on the aperture disk. The objective lens is made of lightly colored glass that has numerous bubbles, striations, and rough edges typical from this time period. The focal length of this lens, combined with the ink markings indicating how far to pull out each draw tube, indicates that this is a Galilean style telescope that gives upright images with a magnification of about 10x.

John Wilkins

The Discovery of a World in the Moone

London, England, 1638 QB41 .W67 1638

An English clergyman, John Wilkins (1614–1672) was one of the founders of the Royal Society and likely the only person ever to head a college at both the University of Oxford (Wadham College, 1648) and the University of Cambridge (Trinity College, 1659). Wilkins was a noted tinkerer who aimed to improve devices for plowing, draining mines, and various other mechanical gadgets. He wanted to spread scientific knowledge and was a great supporter of the Copernican system. Inspired by the lunar details one could observe with the telescope, Wilkins tried to show similarities between the Moon and the Earth; for example, Proposition 11 states: "as their world is our Moone, so our world is their Moone." He concludes that just as the Earth is populated by inhabitants, so too is the Moon. The frontispiece alerts readers to his purpose, with support from the Roman philosopher Seneca, in declaring his certainty that in comparison to God, all else is limited. The full text appears on the Internet at the Project Gutenberg site.

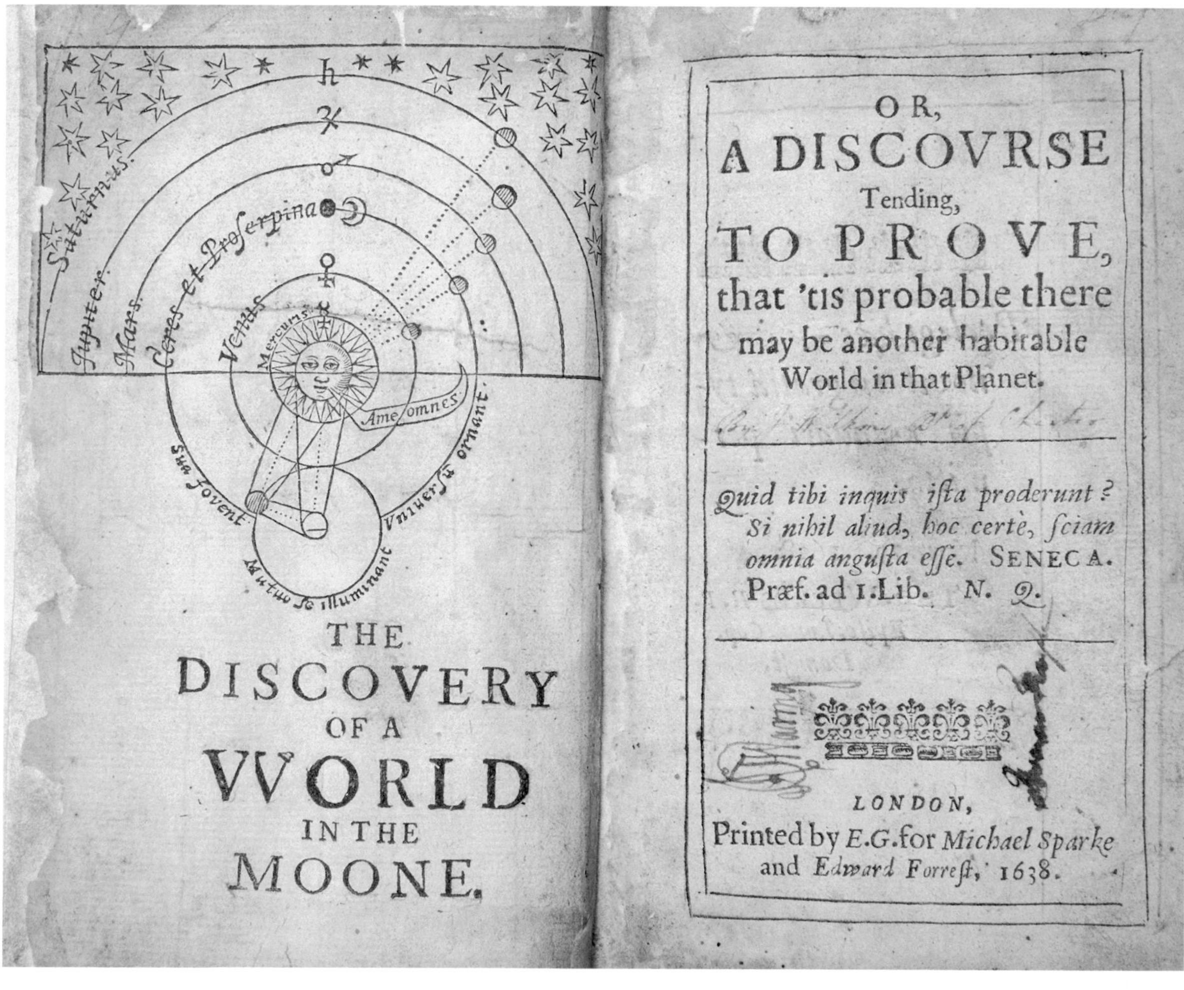
Saturnus.
Jupiter.
Mars.
Ceres et Proserpina
Venus
Mercurius
Ame omnes
Sua fovent
Universu ornant
Mutuo se illuminant
THE
DISCOVERY
OF A
WORLD
IN THE
MOONE.
OR,
A DISCOVRSE
Tending,
TO PROVE,
that 'tis probable there
may be another habitable
World in that Planet.
Quid tibi inquis ista proderunt? Si nihil aliud, hoc certè, sciam omnia angusta esse. SENECA. Præf. ad 1.Lib. N. Q.
LONDON,
Printed by E.G. for Michael Sparke
and Edward Forrest, 1638.

Carlo Manzini *L'occhiale all'occhio* **Bologna, Italy, 1660 TP857 .M29 1660**

By 1660, the production of lenses needed for telescopes and microscopes involved many steps carried out by trained workers using specialized tools. Tubes and their decorative features might be subcontracted, but the techniques for transforming glass into lenses involved innovation, expertise, and secrecy, with close supervision from the master craftsman. The workshops of Eustachio Divini (1610–1685) and Giuseppe Campani (1635–1715) in Rome became the most famous, earning large commissions from wealthy patrons and royalty. Whereas the quality of tubes was certainly important, it was the optical performance of the lenses that built reputations. In the 1660s, Divini and Campani even engaged in competitions to demonstrate the superiority of their lenses. Over the course of these contests, the need to develop test patterns to prevent cheating led to targets with random letters placed in rows of decreasing font size; these became the forerunners of the modern eye chart.

In his *L'occhiale all'occhio,* Carlo Manzini (1600–1677) described and illustrated the lens production process in great detail: how to select glass for the lenses, how to cast and turn (on a lathe) carefully designed molds used for grinding them, and how to select and apply the special abrasives needed to polish them. Here, a workman holds a lens in his left hand, carefully shaping it by holding it against a special lathe powered by his right hand.

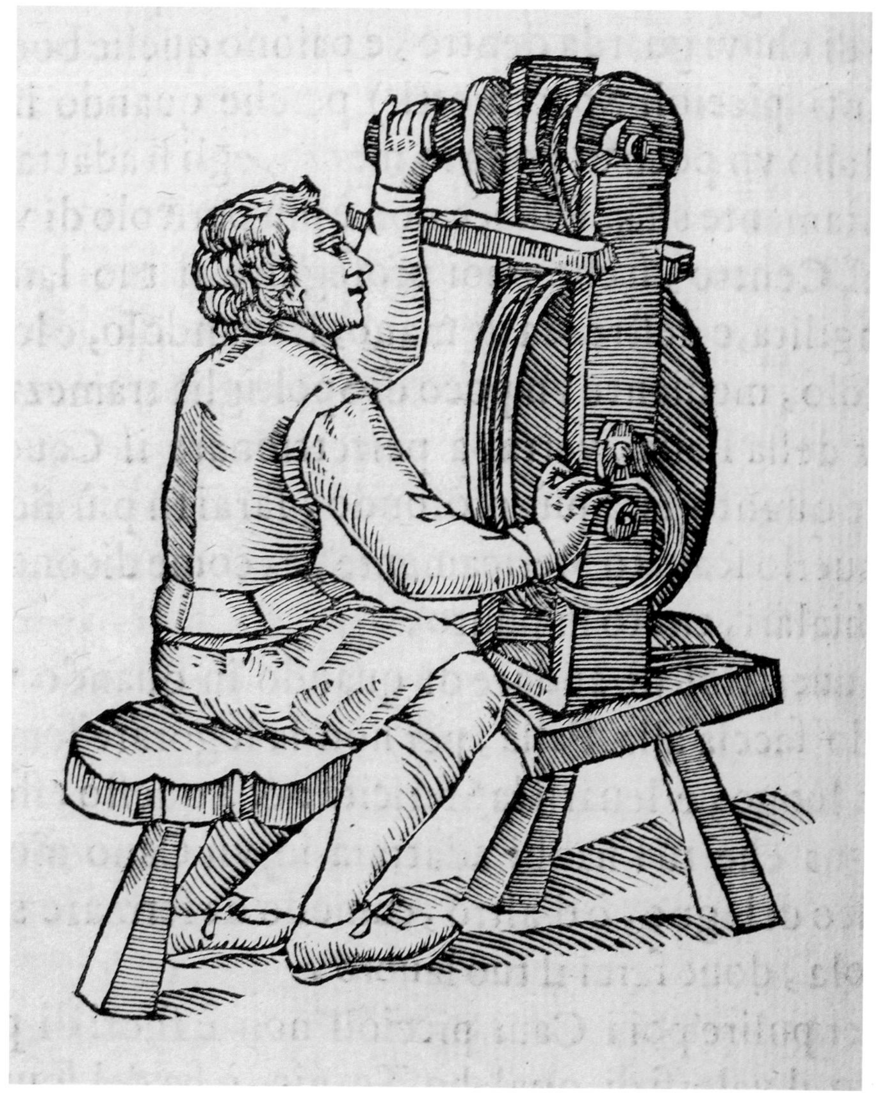

Unidentified Maker

Refracting Telescope

Germany, c. 1660 M-423

In the 16th and 17th centuries, European nobles and royals were often encouraged to learn the art of turning, that is, of using a lathe. Russian Czar Peter the Great, for example, was said to be exceptionally skilled. Some of the most exquisitely turned artifacts are made out of ivory, but this example stands out. Of the handful of ivory telescopes known today, this is the largest and by far the most spectacular.

Likely made around 1660 out of elephant ivory, this turned, tapered, ornate refracting telescope has raised bands of ivory in a scallop cut. It has five ivory draws with ivory ferrules, each of which unscrews from its tube, and a brass erector tube in the fifth draw. Its original objective lens and three eyepiece lenses provide a clear, upright image. Despite the fragility of ivory, as seen in a few small cracks in this piece, this telescope is in extraordinarily fine condition and is one of the finest examples of craftsmanship of any object in the Adler collections.

Christiaan Huygens

Systema Saturnium

The Hague, the Netherlands, 1659
QB671 .H9 1659

Dutch mathematician, physicist, and astronomer Christiaan Huygens (1629–1695) benefited greatly from his exposure to scientific ideas from an early age. His father, Constantijn, was a friend of René Descartes, whose ideas influenced Huygens considerably. Christiaan wrote an important analysis on the nature of light and built numerous telescopes, many of which are now on display at Leiden's Museum Boerhaave in the Netherlands. Using one of his own telescopes, he discovered the first known moon of Saturn, Titan, and observed the Orion Nebula.

Along with illustrations of these discoveries, he provided the first thorough, and ultimately correct, explanation of the rings of Saturn in his landmark volume, *Systema Saturnium,* which is shown here. He first recognized that the changing inclination of these rings would produce different observations of the bulges surrounding Saturn. Huygens was also the first to identify the concept of "seeing," the notion that atmospheric conditions play an important role in the quality of telescopic observations. Huygens also developed techniques for improving the accuracy of a pendulum clock, and he tried to make a pocket watch. In 1666, Huygens took a position at the Paris Observatory, where he remained until 1681, when he returned to his native country. By then, telescopes had increased greatly in length. In response, Huygens developed the tubeless or aerial telescope, and even published a fascinating book on the possibility of extraterrestrial life.

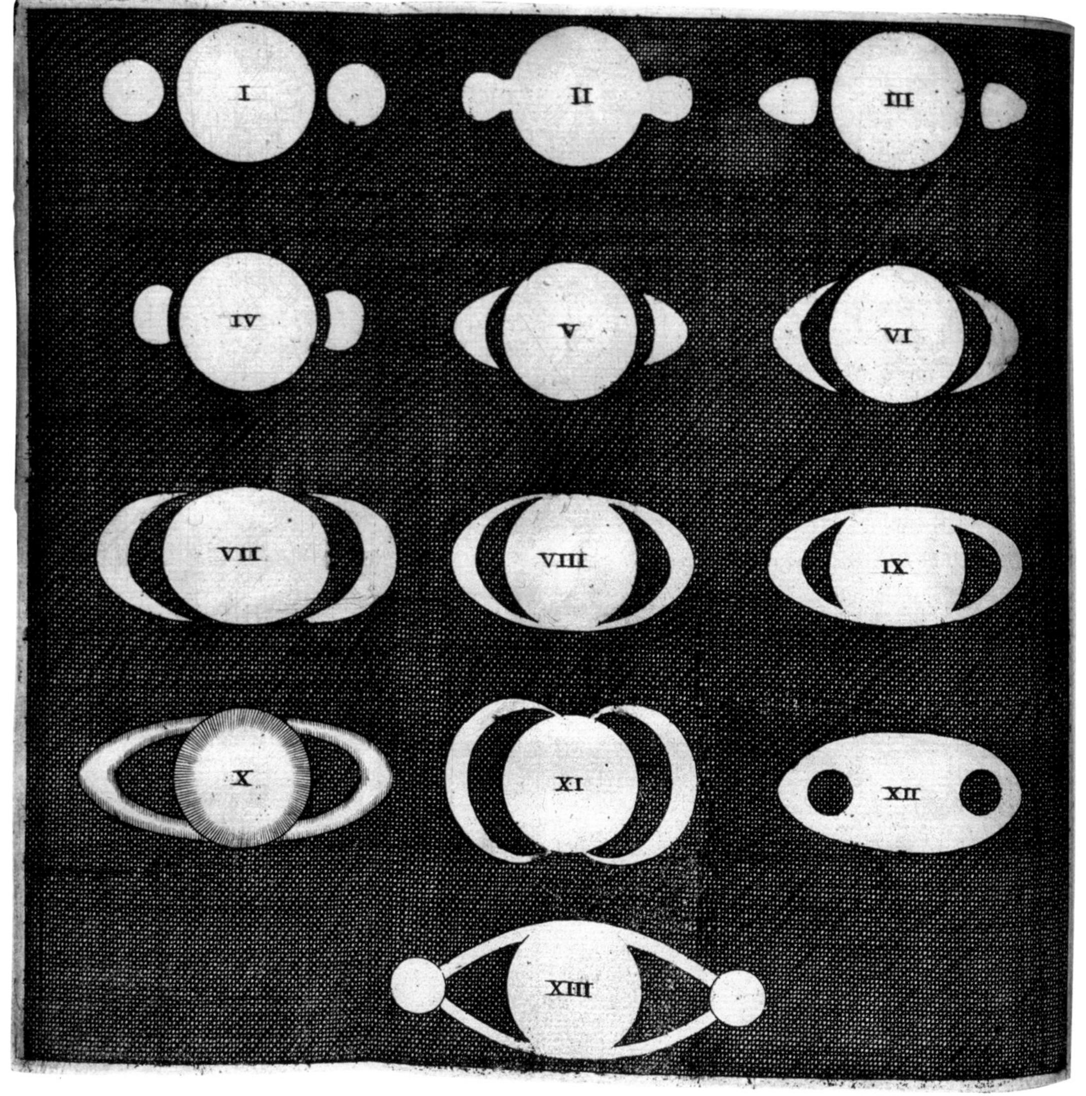
I
II
III
IV
V
VI
VII
VIII
IX
X
XI
XII
XIII

Unidentified Maker, "AG"

Refracting Telescope

France, c. 1675 M-422

This French refracting telescope, dating from ca. 1675, features a main tube covered in dark red Moroccan leather. The leather has gold fleur-de-lis tooling and decorative border motifs, with green vellum at either end. There are two slip-on end caps covered in the same leather and decoration; they are supposed to protect the lenses at each end of the telescope, but the objective lens does not survive. Despite its extensive decorative elements, this instrument has seen a lot of use. It includes three green-vellum-covered draws and an eyepiece tube with an ebony mount for the eyepiece, but it is missing one of the three lenses of the eyepiece unit. One end cap shows the initials "AG". The existence of at least three other telescopes with the same tooling and initials (two in the Netherlands and one in New Zealand) implies that "AG" indicates the name of the maker and not of the owner.

Bernard le Bovier Fontenelle

Entretiens sur la pluralité des mondes

Paris, France, 1686 QB54 .F6 1686

The long-lived Bernard le Bovier Fontenelle (1657–1757) wrote many works about philosophy and science, especially in defense of the worldview put forward by his fellow but earlier Frenchman, René Descartes. Fontenelle's most important work, *Conversations on the Plurality of Worlds*, illustrated here, appeared first in French, not in Latin, the usual choice of scholarly work. *Conversations* reached popular audiences in several editions and in numerous languages. While managing to avoid the prohibition of the Catholic censors, Fontenelle made clear his defense of the Copernican Sun-centered system and his support of the possibility of life elsewhere in the Universe.

The conversations in this book take place between a philosopher and an enlightened woman, a marquise, who shows herself to be more than his equal in her ability to understand the new enlightened, scientific worldview and the consequences of the Copernican system. For example, it seemed plausible that each star, like the Sun, is surrounded by planets inhabited by intelligent beings. In looking at this theatrical frontispiece, one cannot help wondering about all of the possible life forms acting out their lives on an infinite number of celestial stages.

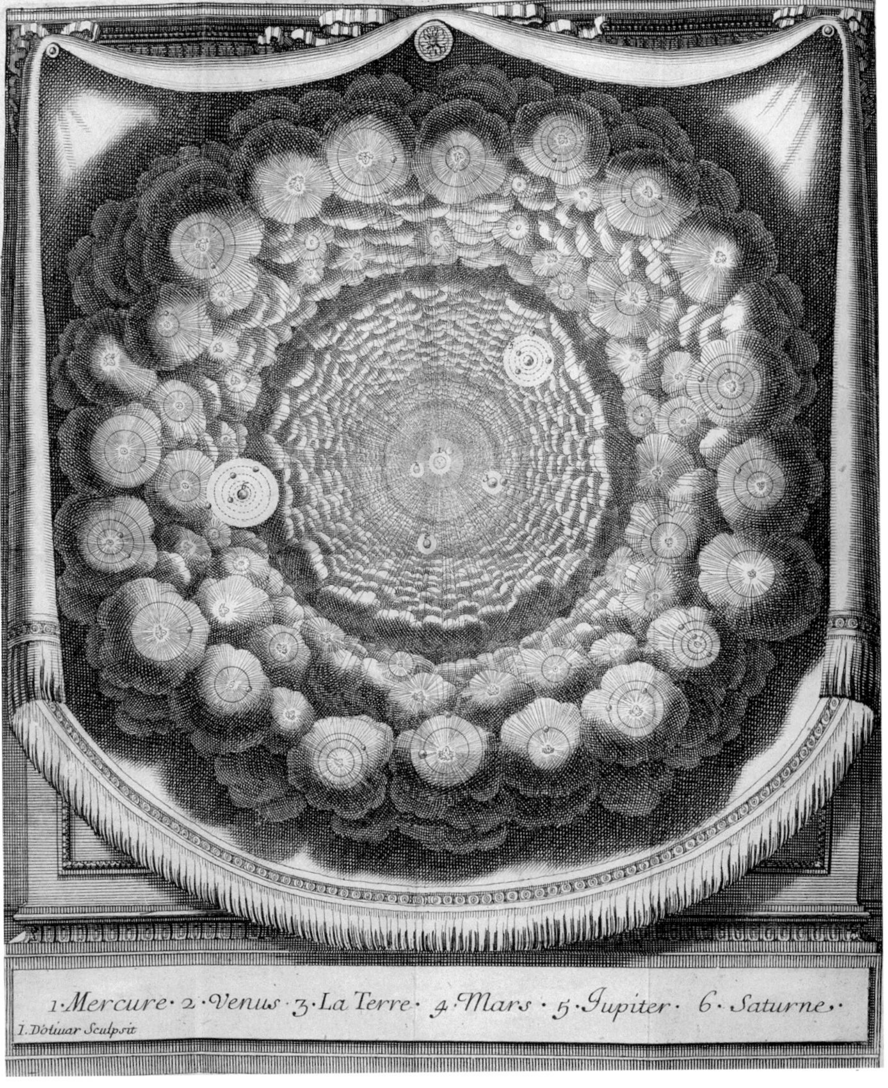
1·Mercure· 2·Venus· 3·La Terre· 4·Mars· 5·Jupiter· 6·Saturne·
I. Dolivar Sculpsit

Philippe-Claude Le Bas **Refracting Telescope** **Paris, France, 1669–1676 W-179**

This 17th-century French refracting telescope, the Adler's oldest complete telescope, includes objective and all tubes and eyepiece lenses. Based on the signature "Le Bas" etched on the objective glass, it is ascribed to Philippe-Claude Le Bas (fl. 1669–1676), optician to the King of France. He made very fine lenses and worked with another optician, Guillaume Ménard, on the Quay de l'Horloge. After his death, his widow – and later his son – carried on the family business. He also made M-186, a large telescopic quadrant in the Adler collections, signed "divisé par Le Bas."

After Christiaan Huygens moved to Paris in 1666, he visited Le Bas several times to find out more about his improved method of polishing lenses. After a year of contact, Christiaan could provide his own brother with only a brief description of this better technique. After the death of Le Bas, Christiaan was unable to obtain the secret from either widow or son.

The main tube of this instrument is covered in black Moroccan leather, with extensive gold-tooled bands of a lace pattern dividing the tube into five decorative panels. The gold tooling extends to the leather ferrules on the four draws, each of which is covered with green vellum. Ink marks on each tube indicate how far to pull out each draw to provide an image in focus; altogether, it extends to about five feet long, and provides a magnification of 20x.

Erhard Weigel *Speculum Uranicum* **Jena, Germany, 1661 QB721 .W33 1661**

Erhard Weigel (1625–1699), a professor of mathematics at the University of Jena, attempted to recast the classical constellations in terms of contemporary European symbols. In his unsuccessful scheme, Ursa Major became "The Elephant of Denmark," and Cygnus became "The Swords of Saxony." His efforts as a popularizer of science included the building of a celestial globe in 1661; still surviving in Dresden's Mathematisch-Physikalischer Salon, it includes tiny holes representing stars and a small opening near the south pole for a viewer to see the illuminated constellations. It served as one of the models for the Adler's Atwood Sphere (1913). Near the end of his life, he met astronomer and telescope-maker Christiaan Huygens.

The right-hand image portrays a school ("Colleg. Tenerife") outfitted with a rooftop observatory where students receive astronomical instruction. In the foreground, an astronomer (perhaps Weigel himself) displays various instruments, including several quadrants, an armillary sphere, and a globe. As described in the shield at the upper right, he views "a prominent comet of 1661," now known as Comet Ikeya-Zhang. It appears at the head of the constellation Aquila the Eagle and just below the arrow of Sagittarius the Archer.

In both images, the astronomer's left hand holds a telescope, his right a banner declaring "videbo coelos" (Latin for "I will see the heavens") and a reference to Psalm 8. Two references are possible for what the astronomer might see: the glory of God ("Lord, our Lord, how majestic is your name in all the Earth! You have set your glory above the heavens", vs. 1) and the insignificance of humans ("When I consider your heavens, the work of your fingers, what are mere mortals that you are mindful of them, human beings that you care for them", vs. 3–4). Some 20 years later, Weigel raised controversy by attempting to use geometric principles to demonstrate the principle of the Divine Trinity.

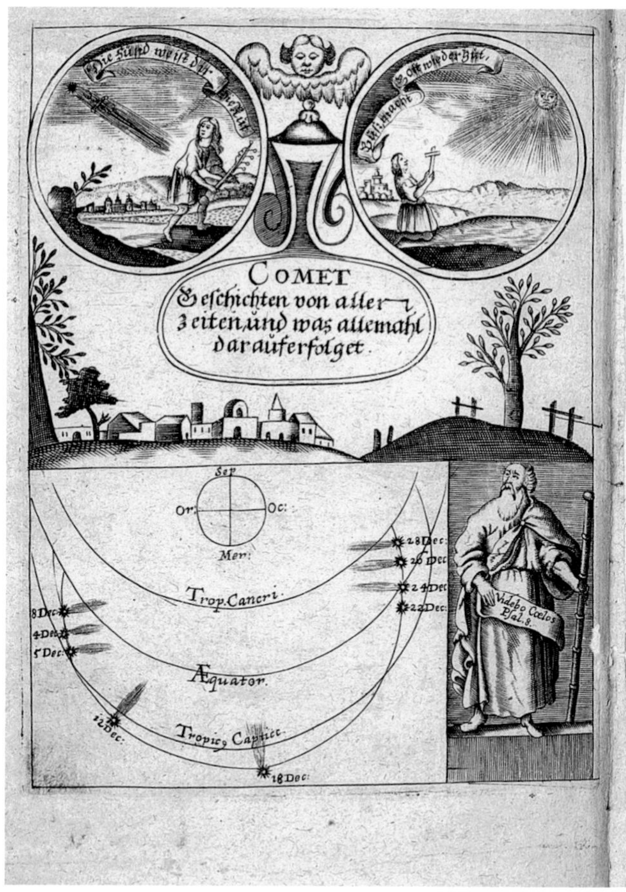
COMET
Geschichten von aller-
zeiten, ünd was allemahl
daraüferfolget.
Sep
Or:
Oc:
Mer:
28 Dec:
26 Dec
24 Dec
22 Dec:
8 Dec
4 Dec
5 Dec
Trop. Cancri.
Æquator.
Tropic. Capric.
12 Dec:
18 Dec:
Videbo Coelos Psal. 8.

Erhard Weigels Math. P.P.
Zur betrachtung des Himmels an führender Comet. Anno 1661.
Colleg. Jenense
Videbo Coelos Psal. 8.
Johann Dürr sculpsit.
In verlegung Thom: Matth: Götzens.

Unidentified Maker **Refracting Telescope** **Italy, c. 1675 M-422a**

While lengthy by modern standards, this Italian refractor, measuring just over 10 feet long, was not at all unusual for its time. The main tube is covered in dark brown leather and is divided into four long rectangular panels by decorative gold tooling with an ornate gold-tooled decoration at each center. Each of the seven draw tubes is covered with a mauve, red, and yellow decorated paper, likely Italian in origin. Each draw tube has two or three markings to indicate how far to pull out the tube so that the image will be in focus. The objective lens is missing, as is the first of the three eyepiece lenses. The innermost tube holds the eyepiece lenses and has writing in Italian to indicate how far that tube should be pulled out when looking at an object very near ("oggetto Vicinissimo") or very far ("oggetto Lontanissimo").

Vincenzo Maria Coronelli

Man looking through a telescope, from *Epitome Cosmographia*

Venice, Italy, 1693 P-167

Vincenzo Maria Coronelli (1650–1718), an Italian Franciscan monk and cartographer, made some of the largest and most spectacular terrestrial and celestial globes of the day, including a pair more than 12 feet in diameter for Louis XIV. Founder of the world's first geographical society, his name continues on through *The International Coronelli Society*, which began in 1952 as a forum for the study, cataloguing, conservation, and restoration of historic terrestrial and celestial globes, armillary spheres, and planetaria.

Here, Coronelli depicts a late-17th-century observer using a telescope. Notice the other instruments at his feet, but the primacy of the telescope is clear. The attached quadrant with suspended weight helps to determine the coordinates of a celestial object, and to map it accurately for a celestial globe or atlas.

John Marshall

Refracting Telescope

London, England, c. 1690 M-424

Another solution addressing the problems of 17th-century telescopes took a rather surprising turn. To make the field of view larger than the very tiny field of a Galilean-style telescope, a monk named Anton Maria Schryle (of Rheita, Bohemia, now in the Czech Republic) designed a telescope in 1645 featuring eyepiece lenses larger than the objective lens. This also resulted in a tube noticeably larger nearer the eye end than at the far or objective end. Such a "reverse-tapered" telescope is very hard to focus, but it provided a popular format from about 1675–1725, especially in England.

John Marshall (1663–1712) made many such instruments, as well as spectacles and microscopes. This elegant example features his name and address ("IOHN MARSHALL LVDGAT STREET LONDON") along with beautiful gold tooling on the tubes and collars of each of the nine draw tubes. A century or so later, someone completely changed the optics, placing an achromatic objective lens (perhaps made around 1800) on the end that used to hold the eyepiece and putting an eyepiece on the former objective end, although that eyepiece does not survive. The reverse taper has thereby been reversed to a more normal form. Nonetheless, the tooling alone makes this a marvelous and beautiful example of the telescope trade more than 300 years ago.

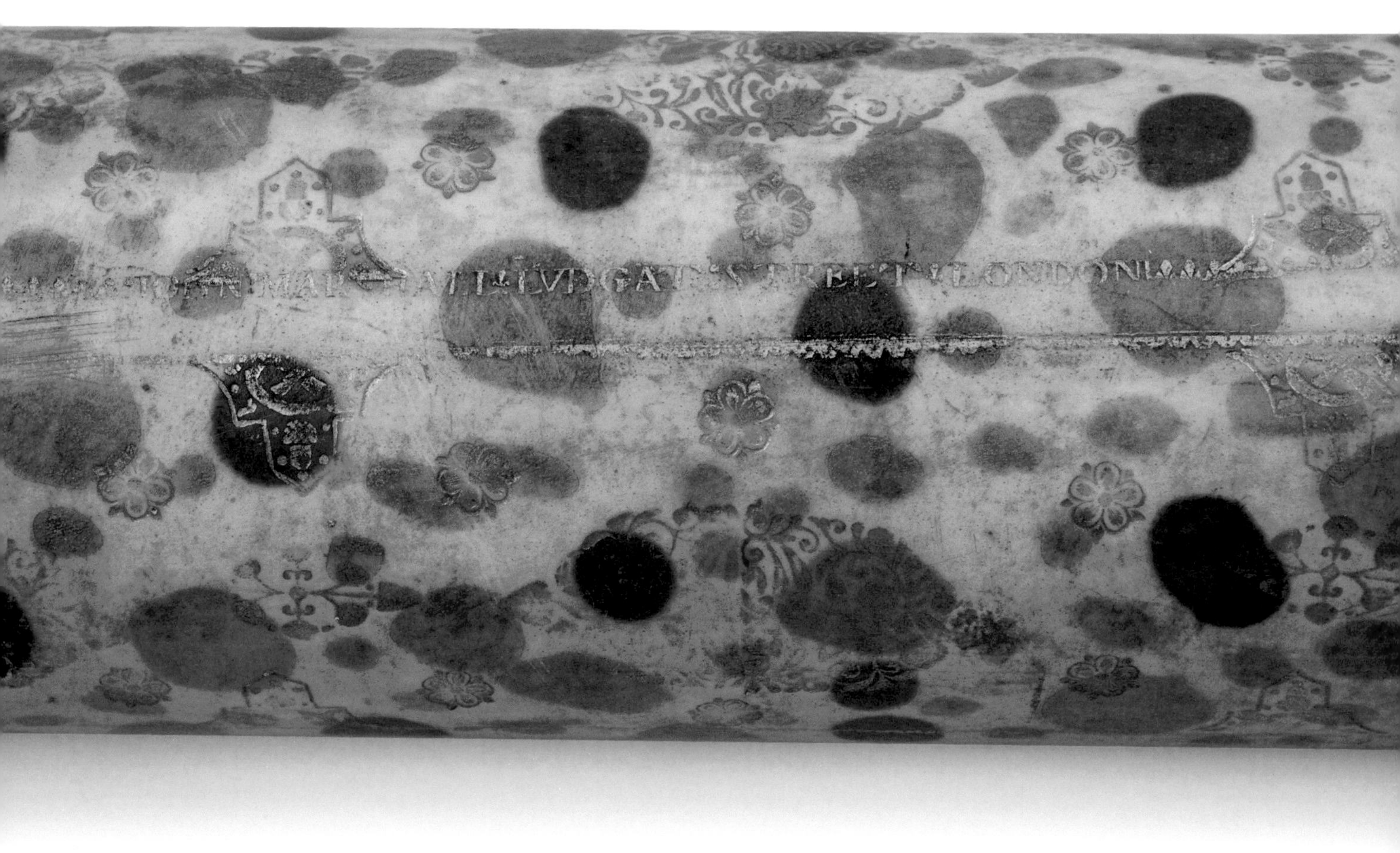
LVDGATE STREET LONDON

Robert Smith *A compleat system of opticks* **Cambridge, England, 1738**

To reduce image blurriness and colored fringes common in mid-17th-century telescopes with higher magnifications, astronomers made objective lenses of longer and longer focal lengths. Giuseppe Campani made usable telescopes for Giovanni Cassini at the Paris Observatory with focal lengths of 100 and 135 feet. Johannes Hevelius built one of nearly 150 feet in length, but Christiaan Huygens claimed useful focal lengths of more than 200 feet, as did James Bradley in London.

The weight of the tube materials causes them to sag noticeably even for 20-foot telescopes, so the mounting of such instruments becomes very important. Hevelius removed much of the tube to reduce weight, but Huygens went further, eliminating the tube altogether, as in this illustration. The objective of such "aerial telescopes" was typically attached to a pole using a moveable, swiveling joint and was located in the dark with a nearby lamp or by a string or rod. The user mounted the eyepiece on a stand, which was moved around (systematically or by trial and error) near the objective's focal point until the resulting image was produced. Telescopes of this length were exceedingly difficult to use, and were seldom used in the 18th century.

When Robert Smith (1689–1768) published his book on optics, he summarized much of what was known about making a refractor or a reflector, and also described what one could see with a telescope. This became a standard resource for 18th-century telescope makers, and provided William Herschel with enough background information to become the leading telescope maker and astronomical observer of his time.

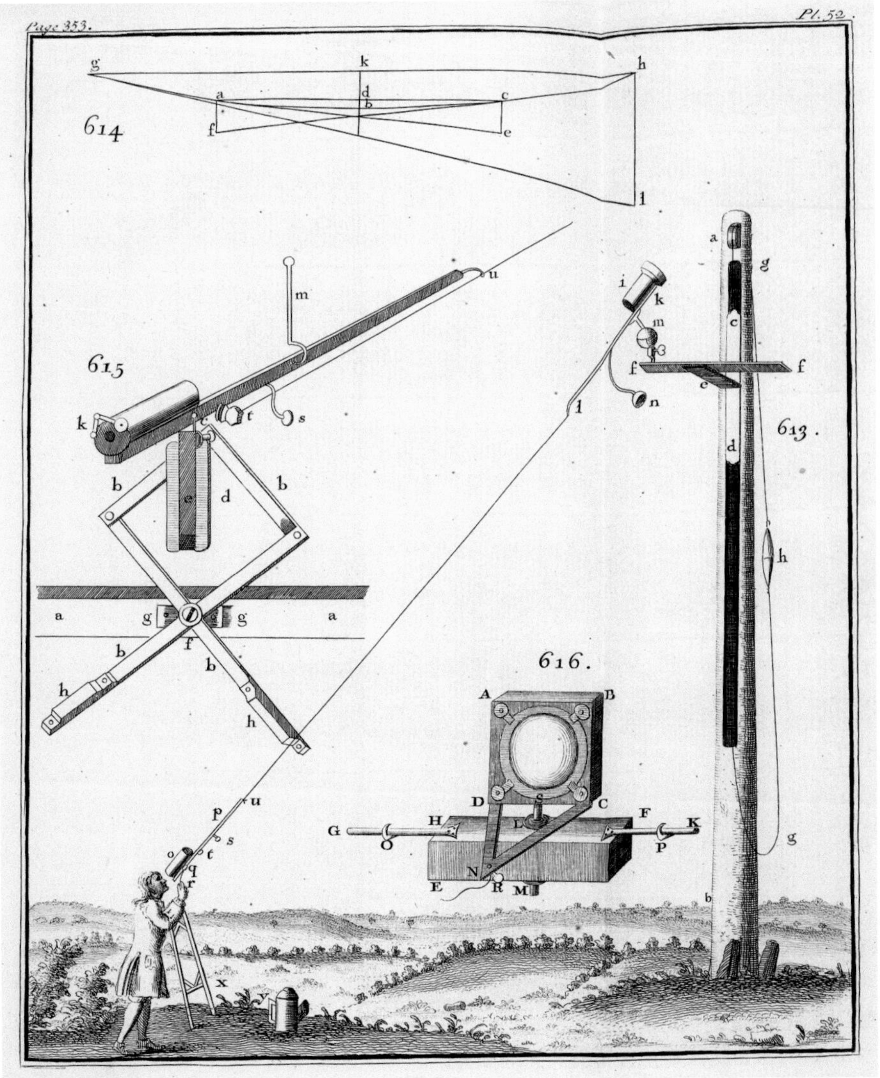
Page 353.
Pl. 52.
614
615
616.
613.

Unidentified Maker

Refracting Telescope

[Italy], ca. 1685 M-428a

Because spherically shaped lenses do not focus all colors at the same distance, images in 17th-century telescopes suffered from colored halos. After 1650, telescopes began to lengthen, reaching 10 or more feet long for more common users and more than 100 feet long for dedicated astronomers, to counter the effects of this chromatic aberration. This late-17th-century refracting telescope, the second-longest in the Adler's collection, measures about 20 feet long when fully extended.

The fractured appearance of the main tube, covered with gold-tooled green vellum, is likely the result of the telescope having been moistened and dried, perhaps from an unsuccessful restoration effort. The gold tooling divides the main tube into three panels, each decorated with a fan motif, a dogtooth roll, a six-pointed star, and other designs. The nine draw tubes are made with wood covered with Italian marbled paper. The last draw tube, likely a replacement, shows construction inferior to the other tubes. A printed fragment of Psalm 71 from a 1702 edition of the Anglican Book of Common Prayer has been pasted on the inside of this tube, indicating that it was repaired after this date.

The style of the telescope strongly resembles several made by Giuseppe Campani (1635–1715), a well-known Italian telescope maker, though the lack of his trademark signature on the greenish yellow objective lens raises questions about this claim. Even so, the holes at the end of the main tube (to allow airflow when draws are moved) are not commonly seen on non-Campani telescopes. This example also features an unusual eyepiece construction. The observer can use the present 1-lens eyepiece to make a Keplerian telescope for viewing celestial objects (upside down); originally, the telescope also came equipped with a 2-lens "converter" for viewing terrestrial objects (right side up). This did not become a common feature of telescopes for more than a century, when brass telescopes included a "night or day" feature, and it is rarely found on non-Campani telescopes this large.

Johann Zahn *Oculus artificialis teledioptricus sive telescopium* **Nuremberg, Germany, 1702**

A Catholic priest from Würzburg, Germany, Johann Zahn (1631–1707) developed great interest and expertise in the camera obscura and the magic lantern, devices that use lenses to project images for public audiences. In his *Oculus artificialis*, Zahn provides numerous descriptions and copperplate illustrations of these and other projection devices and suggests the use of a lens cover, a key step in the development of the photographic camera.

Zahn carefully studied the function of the eye, the process of vision, and the use of lenses and mirrors. Using the principle of the camera obscura, enhanced with a telescope and a combination of lenses and mirrors that enlarged images, Zahn made his own solar observations and showed his readers how to do so as well.

This unusual illustration provides rare details not only of the telescope itself but also of a contemporary mount supporting it. Note the great length of this telescope and the hand crank and screw used to raise and lower it. On the lower left, two ladies discuss the operation of the eye and lenses while the *putti* below try their hands at smaller telescopes and at grinding lenses. On the right, traditional symbols of astronomy (armillary, quadrant, and astrolabe) are shown as having been eclipsed by the importance of the telescope. At the top, even the divine eye seems to look down from heaven with the aid of this device.

OCULUS ARTIFICIALIS
Teledioptricus
IOANNIS ZAHN.
FUNDAMENTUM III
Mechanicum seu Practicum
FUNDAMENTUM II,
Mathematico - Diopttricum.

Nicolas Bonnart **"L'Astrologie"** **Paris, France, c. 1690 P-161**

Nicolas Bonnart (1636–1718), a noted illustrator of fashion and costume, often included objects in his portraits of people and ideas. Here, astrology personified is portrayed in elaborate clothing. She holds the symbol of astronomy, an armillary sphere, in her left hand, surrounded by objects associated with learning, while a drawing instrument rests beneath her right hand. A globe in the background increases the sophistication of the scene. The centerpiece is the telescope, elaborately decorated and resting on a stand far more ornate than necessary.

The text reads: "L'astrologie / Je fais par mon pouvoir tout ce que je souhaite; J'attire à mon party tous les gens curieux, qui par mes Instruments et ma longue lunette / Découvurent avec plaisir le mouvement des cieux. / Chez N. Bonnart, rue St. Jacques à l'aigle, avec privil." "Astrology. I do with my ability all that I wish; I attract all curious people, who with my instruments and my long telescope discover with pleasure the movement of the skies. At N[icolas] Bonnart, St. Jacques Street at the eagle, with privil[ege]."

R·B·del
N.B. Sc.

Unidentified Maker

Refracting Telescope

Italy, c. 1675 M-425

Although the instrument is unsigned, the tooling on the main tube and the style of paper marbling strongly resemble examples known from Italy. The main tube, covered in brown leather, is rich with gold tooling in eight (4 x 2) rectangular sections. The five draw tubes have ferrules gold-tooled in a manner similar to the main tube. The draws are covered in 17th-century Italian marbled paper; they may have been re-covered, as the papering goes up to, but not under, the ferrules. The telescope has only one surviving lens, the first one of the 3-lens eyepiece, which has dark horn fittings. Each tube has several extension marks, suggesting that it may have had more than one set of lenses over its lifetime. With an extended length of more than six feet, it would have needed a sturdy mount to hold it steady.

Franciscus Philippus Florinus

"Sternwerk" from *Oeconomus prudens at legalis*

Nuremberg, Germany, 1705–1719 P-174

Franciscus Philippus Florinus illustrated a variety of contemporary medical, astronomical, and other scientific scenes, as well as a fourteen-volume series on the management of an estate, from which this illustration is taken. Here, he depicts a group of people making nighttime observations of the sky using a variety of astronomical instruments, including a telescope. This private personal rooftop observatory, typical of a wealthy nobleman, also overlooks his estate.

Unidentified Maker **Refracting Telescope** **Germany, c. 1690 M-427**

A carefully designed and very well-made eyepiece unit can be inserted into the final draw tube in either direction, providing either a larger image or a more distinct image. The hand-written German text on the eyepiece provides instructions for how to do just that. At full extension, this large telescope reaches a length of 2,300 mm, or just over 7.5 feet. The main tube is deep green to dark brown, richly decorated with gold-tooled bands of lions, a floral motif, and five-pointed stars. The six draw tubes are covered in speckled tan paper, with the first five draws having green leather ferrules. One of the ferrules has the same lion and star design as the main tube, while the other ferrules are gold-tooled with the weaving floral pattern. The original objective lens survives, but only one eyepiece lens remains of the three-lens erector tube. The lenses are mounted in horn, with turned wooden caps protecting them. The overall impression is of a solid, well-crafted, attractive instrument with strong Italian decorative themes.

Nicolas de Fer

Illustrations of the Paris Observatory

Paris, France, 1705 P-235

Louis XIV (1638–1715) ruled France for seventy-two years. Known also as Louis the Great, he is best known as the Sun King in honor of the newly accepted Copernican world view; that is, his court and his empire should revolve around the king just as the planets orbit the Sun. Louis, a great patron of the arts and sciences, participated in numerous wars and sought to expand his empire throughout the world. He combined these interests in the construction of the Paris Observatory, founded in 1667 in large part to make astronomical observations needed to determine longitude. Its practical purpose was thus to develop France's maritime power and increase her international trade. The Paris Observatory was completed in 1671; in 1675, English King Charles II followed suit, commissioning the Royal Greenwich Observatory to carry out the same tasks for England.

The Paris Observatory inaugurated the age of great observatories that functioned almost as astronomical temples. Upon completion, this observatory was dedicated with several telescopes, as well as naked-eye instruments. Italy's Giovanni Cassini became the first director and used telescopes by fellow Italian Giuseppe Campani to discover four moons of Saturn and the division in Saturn's ring that now bears his name. Some of the longer telescopes, as illustrated in the sketch at the bottom, were too large to fit inside the observatory and required the use of a water tower to support them.

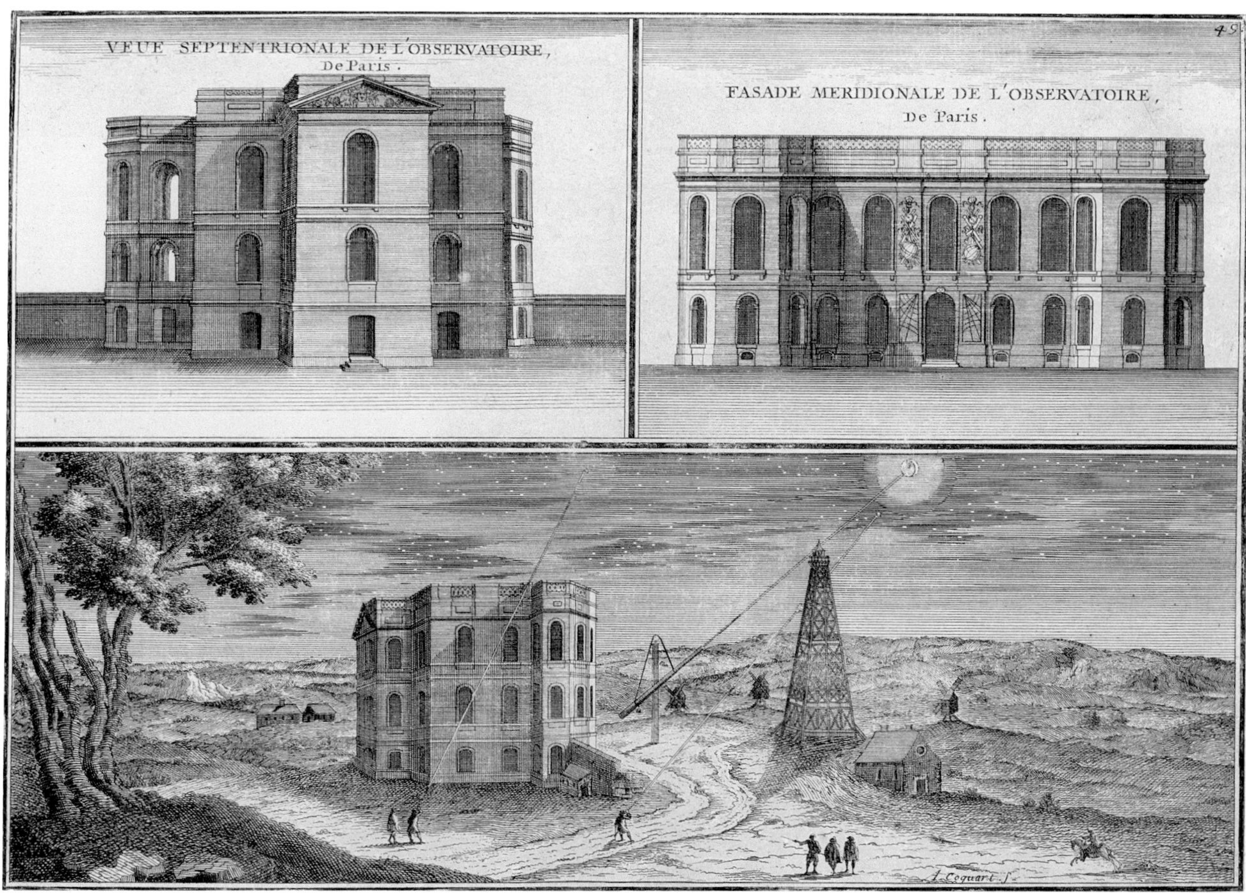
VEUE SEPTENTRIONALE DE L'OBSERVATOIRE,
De Paris.
49
FASADE MERIDIONALE DE L'OBSERVATOIRE,
De Paris.
J. Coquart f.

Pietro Francis de Willevaux

Refracting Telescope

Italy, 1703 M-430

Built in 1703, this refracting telescope features a beautiful ink signature that identifies its maker and date. Inscribed "P. Petrus Franciscus de Willevaux Sacerdos Capuccinus Fecit / Anno 1703", the telescope was made by Father Pietro Francis de Willevaux, an Italian Capuchin friar about whom little is known. Brown Moroccan leather covers the main pasteboard tube, which has elaborate gold tooling of flowers, repeating rectangular panels, and a border pattern at the tube ends. The five draw tubes are covered in green vellum with the same gold tooling, their ends bound with the same brown leather and gold border as the main tube. Although the main objective lens is missing, the three lenses of the eyepiece survive. Much of the green vellum has flaked off from two of the draw tubes. When the telescope is fully extended, it is about 43 inches (1,100 mm) long. It is the only known telescope made by him, though its quality suggests he would have made many more.

P. Petrus Frãn:us de Willeuaux Sacerdos Capuccinus Fecit
Anno .1703.

[Pietro Patroni]

Refracting Telescope

Italy, early 18th century M-460

This unsigned early-18th-century Italian telescope strongly resembles the work of Pietro Patroni (c. 1676–1744). The air release holes at the end of the main tube, as well as the instructions inscribed in Italian on the erector tube, are characteristic of his work (see No. 37). On this eyepiece unit, the instructions on one side read, "Voltando questo Canello in dentro Si vederá Loggetto piú Grande", on the other side, "Voltando questo Canello in dentro Si vederá Loggetto piú Chiaro". As in the other Patroni, for a larger image, insert one way, for a clearer image insert the opposite way. With the "grande" side out, the magnification is 11.7x; with the "chiaro" side out, the magnification is 9.1x.

The main tube is covered in black leather with no tooling. Of the five green vellum draw tubes, the first four have formed-leather ferrules, the fifth a horn ferrule adjacent to an ivory eyepiece mount. The vellum, badly soiled in places by use, has a gold-tooled border near the ferrules. The extension marks are indicated by tooling, suggesting a fairly sophisticated original buyer.

Pietro Patroni

Refracting Telescope

Milan, Italy, c. 1720 M-426

In the hands of a careful maker, telescope tubes fit together so snugly that they are sometimes difficult to slide together because they are air tight. One solution involves placing tiny holes near the objective lens to allow air to escape, as is the case with this example. This well-constructed, large (2,500 mm or over eight-foot-long) early-18th-century Italian refractor comes from the workshop of Pietro Patroni (c. 1676–1744), a famous Milanese maker of monocular and binocular microscopes and telescopes. James Mann (c. 1685–1750), a noted English telescope-maker, once sought to improve his own reputation by inviting the public to compare (favorably, he hoped) his products with those made by the "celebrated Pietro Patroni of Milan."

His signature ("Pietro Patroni in Milano") appears etched around the perimeter of the objective lens. The main tube and draw ferrules are covered with gold-tooled green vellum. The gold tooling, in border patterns of stars and flowers, divides the main tube into four long rectangular panels, with a decorative motif at the center of each panel. The seven draw tubes are covered in multi colored paper of red, yellow, green, mauve, and purple. On the erector tube, instructions appear in Italian: "Voltando' in dentro questo Canello / si vederá L'oggetto piú Grande" and "Voltando in dentro questo Canello / si vederá L'oggeto piú Chiaro"; that is, insert erector tube one way to see the object larger, insert it the other way to see the object more clearly.

Voltando in dentro questo Canello
Si uederà Loggetto più Chiaro
Voltando in dentro questo Canello
Si vederà Loggetto più Grande

Unidentified Maker

Refracting Telescope

France, c. 1700 M-456

The main tube and end caps are covered with artificially pebble-grained leather, stamped with gold tooling of fleur-de-lis and other decorative art. This effect contrasts nicely with the gold-tooled, smooth bright green vellum of the four draws. Turned ebony lens mounts at either end for the objective and eyepiece lenses complete the telescope. The three-lens eyepiece assembly includes a very useful focus adjustment, with the last lens sliding in and out a considerable distance. The small objective lens is cut very ragged on the edges and may be a replacement. In its current configuration, it gives a magnification of 24x.

Domenicus Lusuerg **Broadside Price List** **Rome, Italy, 1698 P-304**

Advertisements, catalogues, and other price lists from even a century ago are uncommon, but those three centuries old are exceedingly rare. The illustration on this example strongly resembles an Adler sector (an early form of calculator, accessioned as SD-3) signed by Domenicus Lusuerg (fl. 1700). On this list, you can find prices for many devices, including sectors, drawing tools, and clocks. The bottom right lists prices for telescopes of different lengths. The sizes range from 10 "palms" (about 30 inches) to 100 palms (about 25 feet), a little more than the length of our second-longest telescope (see No. 28). The longer telescopes on this list were not kept in stock but made to order. Prices are given in *scudi*; although such comparisons are quite difficult to make with any consistency, one *scudo* may be regarded as about one hundred dollars. A large telescope would have been far beyond the reach of anyone other than a wealthy or noble purchaser.

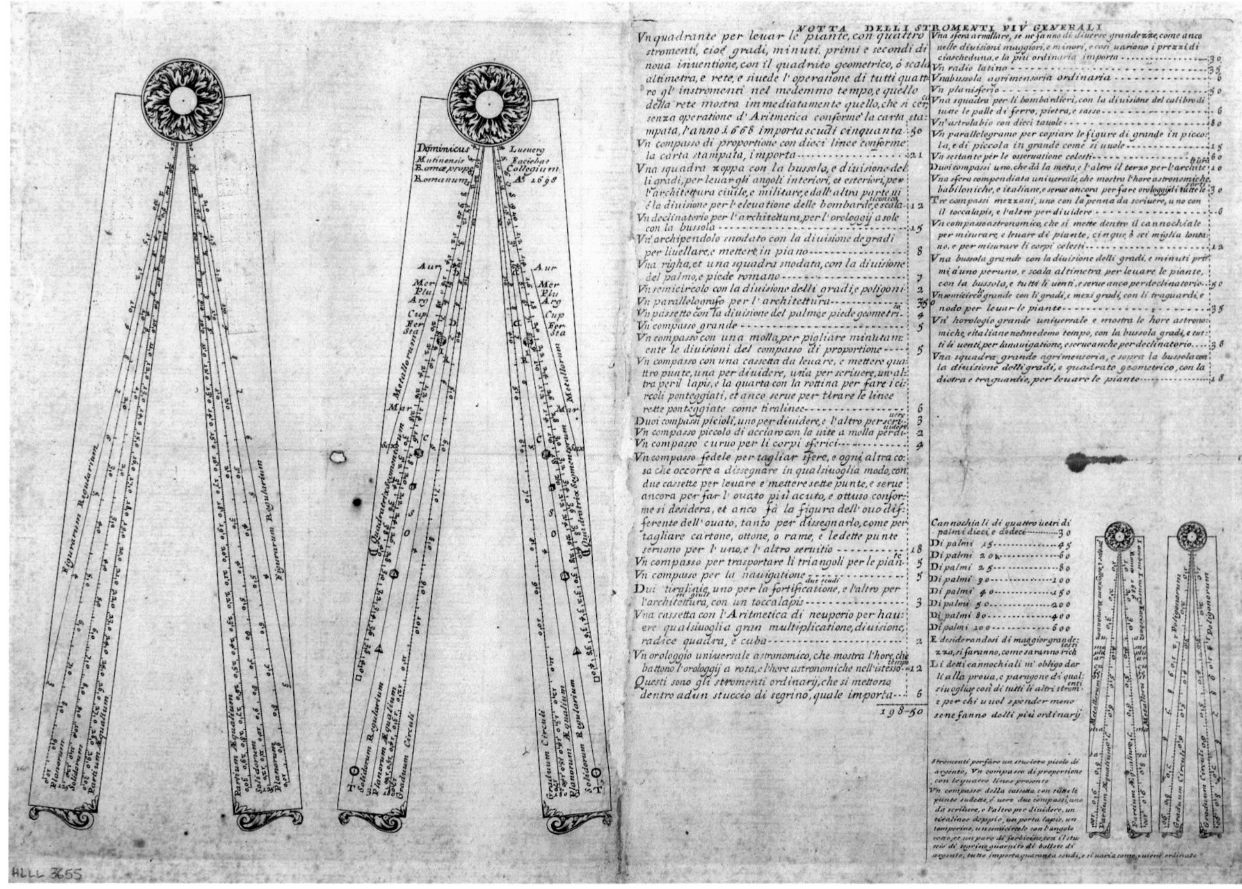

NOTTA DELLI STROMENTI PIV' GENERALI

Vn quadrante per leuar le piante, con quattro stromenti, cioè gradi, minuti, primi e secondi di noua inuentione, con il quadrato geometrico, ó scala altimetra, e rete, e siuede l'operatione di tutti quattro gl'instromenti nel medemmo tempo, e quello della rete mostra immediatamente quello, che si cerca, senza operatione d'Aritmetica conforme la carta stampata, l'anno 1668 importa scudi cinquanta ---- 50

Vn compasso di proportione con dieci linee conforme la carta stampata, importa ---- 21

Vna squadra zoppa con la bussola, e diuisione delli gradi, per leuargli angoli interiori, et esteriori, per l'architettura ciuile, e militare, e dall'altra parte ui é la diuisione per l'eleuatione delle bombarde, e scala ticonica ---- 12

Vn declinatorio per l'architettura, per l'orologgij a sole con la bussola ---- 15

Vn archipendolo snodato con la diuisione degradi per liuellare, e mettere in piano ---- 8

Vna righa, et una squadra snodata, con la diuisione del palmo, e piede romano ---- 7

Vn semicircolo con la diuisione delli gradi, e poligoni ---- 2

Vn parallelografo per l'architettura ---- 3 5 o

Vn passetto con la diuisione del palmo, e piede geometrico ---- 4

Vn compasso grande ---- 5

Vn compasso con una molla, per pigliare minutamente le diuisioni del compasso di proportione ---- 5

Vn compasso con una cassetta da leuare, e mettere quattro punte, una per diuidere, una per scriuere, un'altra per il lapis, e la quarta con la rottina per fare i circoli ponteggiati, et anco serue per tirare le linee rette ponteggiate come tiralinee ---- 6

Duoi compassi picioli, uno per diuidere, e l'altro per scriuere ---- 3

Vn compasso piccolo di acciaro con la uite a molla per diuidere ---- 2

Vn compasso curuo per li corpi sferici ---- 4

Vn compasso fedele per tagliar sfere, e ogni altra cosa che occorre a dissegnare in qualsiuoglia modo, con due cassette per leuare e mettere sette punte, e serue ancora per far l'ouato più acuto, e ottuso conforme si desidera, et anco fà la figura dell'ouo differente dell'ouato, tanto per dissegnarlo, come per tagliare cartone, ottone, o rame, e le dette punte seruono per l'uno, e l'altro seruitio ---- 18

Vn compasso per trasportare li triangoli per le piante ---- 5

Vn compasso per la nauigatione ---- 5

Dui tiralinie, uno per la fortificatione, e l'altro per l'architettura, con un toccalapis due scudi e giulii ---- 3

Vna cassetta con l'Aritmetica di neuperio per hauere qualsiuoglia gran multiplicatione, diuisione, radice quadra, e cuba ---- 2

Vn orologgio uniuersale astronomico, che mostra l'hore, che battono l'orologgij a rota, e l'hore astronomiche nell'istesso tempo ---- 12

Questi sono gli stromenti ordinarij, che si mettono dentro ad un stuccio di segrino, quale importa ---- 6

198-50

Vna sfera armillare, se ne fanno di diuerse grandezze, come anco nelle diuisioni maggiori, e minori, e cosi uariano i prezzi di ciascheduna, e la più ordinaria importa ---- 30

Vn radio latino ---- 35

Vna bussola agrimensoria ordinaria ---- 6

Vn planisferio ---- 50

Vna squadra per li bombardieri, con la diuisione del calibro di tutte le palle di ferro, pietra, e sasso ---- 6

Vn'astrolabio con dieci tauole ---- 80

Vn parallelogramo per copiare le figure di grande in piccola, e di piccola in grande come si uuole ---- 15

Vn sestante per le osseruatione celesti ---- 60

Duoi compassi uno, che dà la meta, e l'altro il terzo per l'architettura ---- 10

Vna sfera compendiata uniuersale, che mostra l'hore astronomiche, babiloniche, e italiane, e serue ancora per fare orologgij di tutte le sorti ---- 30

Tre compassi mezzani, uno con la penna da scriuere, uno con il toccalapis, e l'altro per diuidere ---- 6

Vn compasso astronomico, che si mette dentro il cannochiale per misurare, e leuare di piante, cinque, ò sei miglia lontano, e per misurare li corpi celesti ---- 12

Vna bussola grande con la diuisione delli gradi, e minuti primi a uno peruno, e scala altimetra per leuare le piante, con la bussola, e tutti li uenti, e serue anco per declinatorio ---- 50

Vn semicircolo grande con li gradi, e mezi gradi, con li traguardi, e nodo per leuar le piante ---- 35

Vn horologio grande uniuersale e mostra le hore astronomiche, e italiane nel medemo tempo, con la bussola gradi, e tutti li uenti, per la nauigatione, e serue anche per declinatorio ---- 36

Vna squadra grande agrimensoria, e sopra la bussola con la diuisione delli gradi, e quadrato geometrico, con la diotra e traguardio, per leuare le piante ---- 18

Cannochiali di quattro uetri di palmi dieci, e dodeci ---- 30

Di palmi 15	45
Di palmi 20	60
Di palmi 25	80
Di palmi 30	100
Di palmi 40	150
Di palmi 50	200
Di palmi 80	400
Di palmi 100	600

E desiderandosi di maggior grandezza, si faranno, come saranno richiesti

Li detti cannochiali m'obligo darli alla proua, e parugone di qualsiuoglia, cosi di tutti li altri stromenti e per chi uuol spender meno sene fanno delli più ordinarij

Stromenti per fare un stucietto picolo di argento, Vn compasso di proportione con le quattro linee presente

Vn compasso della cassetta con tutte le punte sudette, é uero due compassi, uno da scriuere, e l'altro per diuidere, un tiralinee doppio, un porta lapis, un temperino, un semicircolo con l'angolo retto, et un paro di forbicine, con il stuccio di segrino guarnito di bollete di argento, tutte importa quaranta scudi, e si naria come saranno ordinate

Unidentified Maker

Refracting Telescope

Italy, 18th century M-458

This elegant telescope illustrates many of the difficulties in identifying historic artifacts, such as telescopes. This example includes no signature and has lost all of its optical elements. The main tube and ferrules are covered in red Moroccan leather, gold-tooled with floral motifs and scroll patterns. These features are typical of Italian telescopes of the second half of the 17th century. The three draw tubes are covered with green vellum in a manner often seen in French telescopes of the early 18th-century. Finally, the erector tube is covered with red-and-yellow marbled paper, similar to M-426 (see No. 37), and M-444 (see No. 42). In such cases, more information is needed from diverse experts on the material culture of the period, including details about the leather, tooling, and paper.

Unidentified Maker

Refracting Telescope

Germany, c. 1700 A-364

This very small (125 mm or five inches short) but elegant refracting telescope features an elaborate zigzag pattern on its main tube. Made of silver, it functions more as a valuable decorative item than as a useful instrument. Nonetheless, its original objective and eyepiece lenses provide a distinct image, albeit with a very small field of view. This unusual item has a family resemblance to three small telescopes in a collection in Kassel, Germany, and to another in a decorative arts museum in Berlin.

Unidentified Maker **Refracting Telescope** **Germany, c. 1700 M-444**

Although the optics and the decorative elements of this instrument are not of high quality, the careful construction of the tubes shows some expertise. The unusual decorations make it quite difficult to assign it a place of origin or date. The main tube of the telescope is covered with impressed wallpaper in a floral green-and-gold design. Although marbled paper and various paste papers are commonly used as tube covers, a more textured wallpaper is quite unusual. The inside of the tube is black, cutting down on any stray light that would reduce the quality of the image. There are five draw tubes in total, covered with red, yellow, and purple marbled paper in a smear pattern. With a length of about 1,420 mm, or just over 4.5 feet, and a magnification of 18, this makes a nice instrument for observations or for decorating a study.

Unidentified Maker **Refracting Telescope with Case** **Unknown Location, c. 1725 M-454**

Like other scientific instruments, such as sundials, telescopes can be made of many different materials. This example is made to resemble ivory but consists instead of bone, safely housed in its original wooden case. The wooden tube case is fashioned similar to the telescope, with turned bands at the center and ends. The tube is smooth, except for those turned bands, and does not include any draw tubes. Screw-on caps protect lenses at both ends; the objective lens is missing, while the eyepiece lens might not be original. The objective end has a screw-on, turned wooden cap; the ivory eyepiece end unscrews for a small amount of focusing. The eyepiece end also has a turned finial. Bone tubes are sometimes used for small, pocket-sized telescopes, but one this size is extremely rare.

Unidentified Maker **Refracting Telescope** **Unknown Location, c. 1725 M-459**

The considerable wear and tear on this telescope suggests that it was used quite a bit during its lifetime. The main tube is covered in leather and would have had leather-covered caps on each end to protect the tubes and the lenses (compare with No. 20). As usual, the tubes are made of layers of paper, in this case covered with green vellum that is decorated on one end with gold stamped tooling. Leather ferrules protect each end, enabling the user to pull out each draw to the length indicated by an ink mark at the opposite end. The objective has an ebony-and-ivory mount; the turned eyepiece mount at the opposite end is probably lignum vitae, another hard wood. The objective is missing, but the eyepiece unit has its three original lenses.

In 1645, Bohemian monk Anton Schyrle developed a terrestrial eyepiece with three lenses to create a large, upright image. Almost overnight, nearly every refracting telescope began to use this new format (or a variation of it), doing away with the single lens eyepiece of the Galilean telescope. For the next century, until the development of achromatic objectives, telescopes used three-lens eyepieces.

The simple decoration and overall appearance of this example are typical of French telescopes from this time period. Although it would not have been the tool of an expert user, the expense of this well-made device kept it out of the hands of an ordinary citizen. Instead, it likely found a home in the residence, or even the pocket, of a well-connected gentleman living a comfortable life.

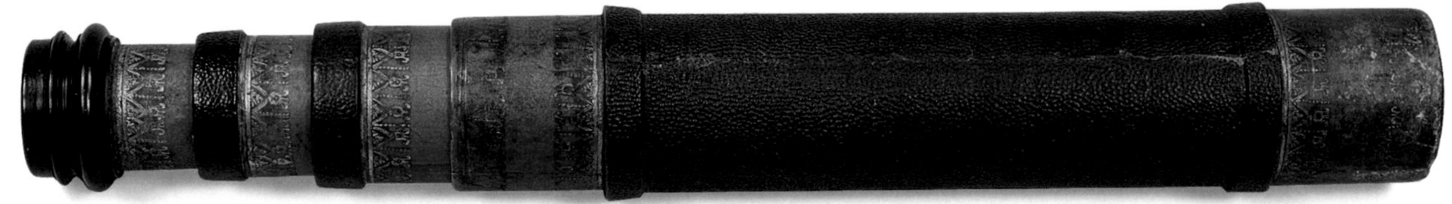

Franciscus Baillou

Refracting Telescope

Milan, Italy, 1738   M-431

This beautiful early-18th-century refracting telescope features a long signature: "Frančus Baillou fecit Mediolani anno 1738", indicating its origins in Milan in 1738. The main tube is covered with brown goat leather, gold tooled with compartments defined by bands of leaves and vertical chains. The bands, and possibly the centerpieces, were made with individual stamps rather than being rolled. Each draw tube is covered with variegated color paste paper, a form of wallpaper, and has a brass ferrule. The objective lens, with its maker's signature etched around its edge, survives in its original olive-wood-and-ivory mount; the eyepiece includes three lenses with ivory mounts.

George le Père

Refracting Telescope

Paris, France, 1744 M-442

Though missing its end caps, this French refracting telescope is otherwise in nearly perfect condition, giving insight into the design and craftsmanship of telescopes about 250 years ago. The maker's signature appears on the objective lens, etched there with a diamond stylus: "George le père quai de l'orloge du palais 1744"; that is, George (the father) had his workshop or sales shop on the Quai de l'Horloge, a bank along the Seine (on the Île de la Cité downstream from the cathedral of Notre Dame) and named for what is now the oldest clock in Paris, installed in 1370 by Charles V. Another telescope (in a private collection) by this maker is signed from the same year. Little is known of this maker, although apparently his son was also involved in the same trade.

The main tube and four draws are covered with a combination of artificially grained dark brown or black leather and gold-tooled green vellum. The gold tooling is in bands, likely individual tools rather than with a roll, in the following styles: short double lines, a longer single pointille line, a leafy flower, and a tuliplike flower. Shagreen is used for most of the main barrel and as the covering for each ferrule on the four draws. The signed objective lens and the three eyepiece lenses have wooden mounts, most likely lignum vitae, one of the hardest, densest woods known and used also for wooden gears in some clocks.

J. G. Rudolph

Refracting Telescope

Dresden, Germany, 1750–1760 M-446

Johann Gottlob Rudolph (1721–1776) was a director of the Mathematisch-Physikalischer Salon in Dresden. There he also made several handsome telescopes, two of which are currently in the Salon's renowned collections, including one that is a stunning reflecting telescope covered in Meissen porcelain.

The illustrated example includes a main tube with four draws and an objective mount with lens and protective threaded-metal cover with slider. The maker's signature, "J.G. Rudolph fecit Dresdn", along with its focal length ("5′" in local units) is engraved on the well-polished objective lens, but the eyepiece does not survive.

The main tube and four draw ferrules are covered with gold-tooled light tan vellum, the circumscribed decoration consisting of stylized geometric and floral motifs. The draw tubes are covered with a blue paste paper that offers a striking contrast with the tan vellum.

Unidentified Maker **Refracting Telescope** **Venice, Italy, c. 1775 M-453**

Glassmaking in Venice has been documented back to at least 982. In the 13th century, regulations addressed the entire production process, how it was made and sold, ranges for prices and taxes, and the relationships between owners, artisans, and other factory workers. Later restrictions included the coming and going of all people to the nearby island of Murano, where factories were located so that the ever-present danger of fire would not burn down the city itself.

The clear, colorless glass known as Venetian crystal (*cristallino*) was invented in the middle of the 15th century. Galileo, like many others, used Venetian glass to build his own telescopes, but in the 17th century, many Murano glassmakers left the island to escape the harsh laws and taxes that impoverished them and moved throughout Europe, spreading their craft.

Even so, by the middle of the 18th century, Venice was well known for its large production of small, inexpensive telescopes, many of them made by several generations of the same family, others by unknown makers. Like most of these devices, this example consists of pasteboard (glued layers of paper); the main tube is covered with leather, the draws with vellum and with horn ferrules. The erector tube, holding the three lenses of the eyepiece, is covered with alum-tawed skin (that is, the flesh side of pig, sheep, or calf, but treated with alum rather than tanned as leather). The objective and eyepiece lenses survive, providing a magnified image of about 10x. The main tube is covered in smooth reddish brown calf leather with rolled tooling at either end of a relatively plain style, one in a lozenge pattern and the other a palmette-and-leaf pattern.

Venetian telescopes similar to this one can be found in numerous collections and even in flea markets.

Thomas Heath

Grand Orrery

London, England, original c. 1740, last expanded c. 1797 DPW-1

Thomas Heath (fl. 1714–1765) owned one of the many shops offering scientific instruments in 18th-century London. He began making instruments c. 1720 and established a family firm that sold devices made by other craftsmen as well, including sundials and surveying tools. Some of the makers who worked for Heath, such as George Adams, later established their own workshops and reputations.

By the early 18th century, the Copernican (Sun-centered) model of the Universe was widely accepted. Astronomers trained their telescopes on the Moon and the planets, looking for evidence that these other worlds might be similar to Earth. Models of the Universe became very popular and took many forms.

This large planetarium is also known as a grand orrery. It shows the planets known from antiquity but displayed in the new Copernican arrangement. Although it does not indicate their relative distances, an elaborate clockwork mechanism moves the planets at their correct relative speeds.

When built around 1740, this orrery included a upraised central ring. It displayed the Julian calendar, which had been replaced in Catholic countries in 1582 but was still used in England until 1750, when the calendar ring had to be altered. After William Herschel discovered Uranus in 1781, a skilled artisan added another outer ring to this orrery to show the new planet, labeled here as "Georgium Sidus" or King George's star. At that time, or perhaps around 1797, the last of the six moons of Uranus featured here were added as well. Of these six, Herschel discovered two and imagined that he saw the other four.

The Orbit of the GEORGIUM SIDUS and his six Satellits The Diameter of this Planet is about

III

Development of the Telescope

...[H]e was looking not only deeper into space but also back in time, and was seeing the evolution of the Universe.

Several strategies led to improvements of the telescope. The small field of view of the simple Galilean telescope gave way to the much-improved image provided by the three-lens eyepiece. The biggest problems in the 17th and 18th centuries remained those of spherical and chromatic aberration. Lens makers then could produce lenses only in a spherical shape, which cannot focus rays at a single focal point. This yields fuzzy images, often with extraneous colors and is labeled as spherical aberration. Just as problematic is a law of physics – rays of different colors do not get refracted equally by lenses, a phenomenon known as chromatic aberration. In practice, these two problems can be difficult to distinguish.

One of the best-known strategies involved making an objective with a long focal length, minimizing the distortions of the image. Another solution uses a mirror rather than a lens to gather the light. The more complete solution to the lens problem was attained in the 1750s with the development of the achromatic lens, which involved two (or three) objective lens components made of different types of glass whose "defects" compensate for each other when the lenses are shaped appropriately.

Isaac Newton had argued that the introduction of colors and blurring was an intrinsic problem of telescopes using lenses and so recommended the use of a mirror to gather light. Several people put their hands to this task, including Robert Hooke and James Gregory. But reflecting mirrors need to be made much more precisely than do lenses, and this proved a very difficult task. Moreover, mirrors until

1850 were made out of speculum metal, a combination of copper and tin. Even under the best conditions, such mirrors reflect about 75% of incoming light, and often much less. There are two basic styles of reflecting telescopes, the Newtonian and the Gregorian. The Newtonian is more convenient for the viewer, allowing the user to stand comfortably; its disadvantage is that the viewer does not look in the direction of the sky. The Gregorian is aimed towards its target, but it sacrifices a great deal of its light-gathering power because of the hole in the primary mirror.

Even so, reflecting telescopes enjoyed popularity because they were so much more compact than refracting telescopes and were therefore much easier to use. A similar benefit marked the popularity of the achromatic lens when it was introduced in the 1750s by John Dollond. The achromatic lens eliminated much of the chromatic aberration that had marked refracting telescopes for more than a century, but at least initially, its chief advantage arose from how it provided achromatic images using a much shorter, and thus more usable, telescope.

In the 17th and 18th centuries, academic and national observatories were mostly preoccupied with measuring and charting the positions of the Sun, Moon, stars, and planets. To carry out these tasks, they attached telescopes to divided circles and quadrants for making precise, quantified observations; there was little interest in the physical properties of the heavens. When William Herschel discovered the planet Uranus, recognizing its novelty not by its movement but by its appearance, he was described by professional astronomers not as an astronomer but as a musician. This undoubtedly had something to do with Herschel's tendency for speculating: in his search for evidence of extraterrestrial life, he claimed that he had no doubt of inhabitants living on the Moon, on Mars, and even on the Sun, and that given the choice between living on Earth or on the Moon, he would not hesitate to choose the Moon. Other convictions led to more significant contributions. He tried to map out the structure of the Milky Way, an important step in furthering our understanding of the cosmos. He also realized that using his powerful reflector, he was looking not only deeper into space but also back in time and was seeing the evolution of the Universe. This insight underlies our contemporary approach to cosmology.

By the end of the 18th century, the development of a middle class, especially in England, created a significant demand for scientific instruments. A telescope in the home, or in the pocket, functioned as an object to look *through*, and often just as importantly, to look *at*. Ornamentation of these devices had long been important, but the mass market now led to the production of much plainer devices as well. Brand names, such as Dollond or Ramsden, could generate higher prices than unsigned pieces and led to fakes and forgeries. Scholars today still debate the authenticity of some pieces signed "Dolon" or "Dolland".

Giambattista della Porta

De refractione optices

Naples, Italy, 1593 Q155 .P66 1593

Giambattista della Porta (1535–1615), an Italian polymath and Renaissance man, devoted his life to the study of science. He is best known for his *Magiae Naturalis* (*Natural Magic*, 1558), which summarizes the diversity of his studies of natural philosophy, including astronomy, astrology, mathematics, meteorology, alchemy, and the occult. His work led to numerous engineering innovations and to the great improvement of the camera obscura, which he described in his *Magiae Naturalis*. Today a few scholars still insist that he had developed the principle of the telescope prior to 1608 and that he had built a model a decade or more earlier. These claims do not have merit, although more investigation may uncover additional evidence to support them.

The earliest known *sketch* of a telescope does come from a letter written by della Porta on August 28, 1609, to Federico Cesi (1585–1630), who is often credited with inventing the word "telescope." More likely, the word, based on Greek roots *tele* ("far") and *skopein* ("to look or see"), was coined by Giovanni Demisiani (d. 1614), a Greek poet, natural philosopher, and mathematician, and a member of the Accadèmia dei Lincei (Academy of Linxes). It was headed by Cesi, who announced the word at a banquet on April 14, 1611, celebrating Galileo's induction to the academy. At this banquet Galileo presented one of his instruments to those gathered there, including Sirtori (see No. 14).

Porta certainly did investigate optics, and he had published some of his findings in 1589. He carried out an extensive investigation of refraction, the bending of light by glass, water, and lenses. He showed, for the first time, that the rays bent by a lens do not all meet in one point, the focus, but that rays entering the lens at the center meet at a point farther from the lens than the rays entering at the edge. This spherical aberration results in blurred images that plagued telescopes until the 1750s. Porta's sketch shown here is the first one ever to illustrate this profound and vexing problem.

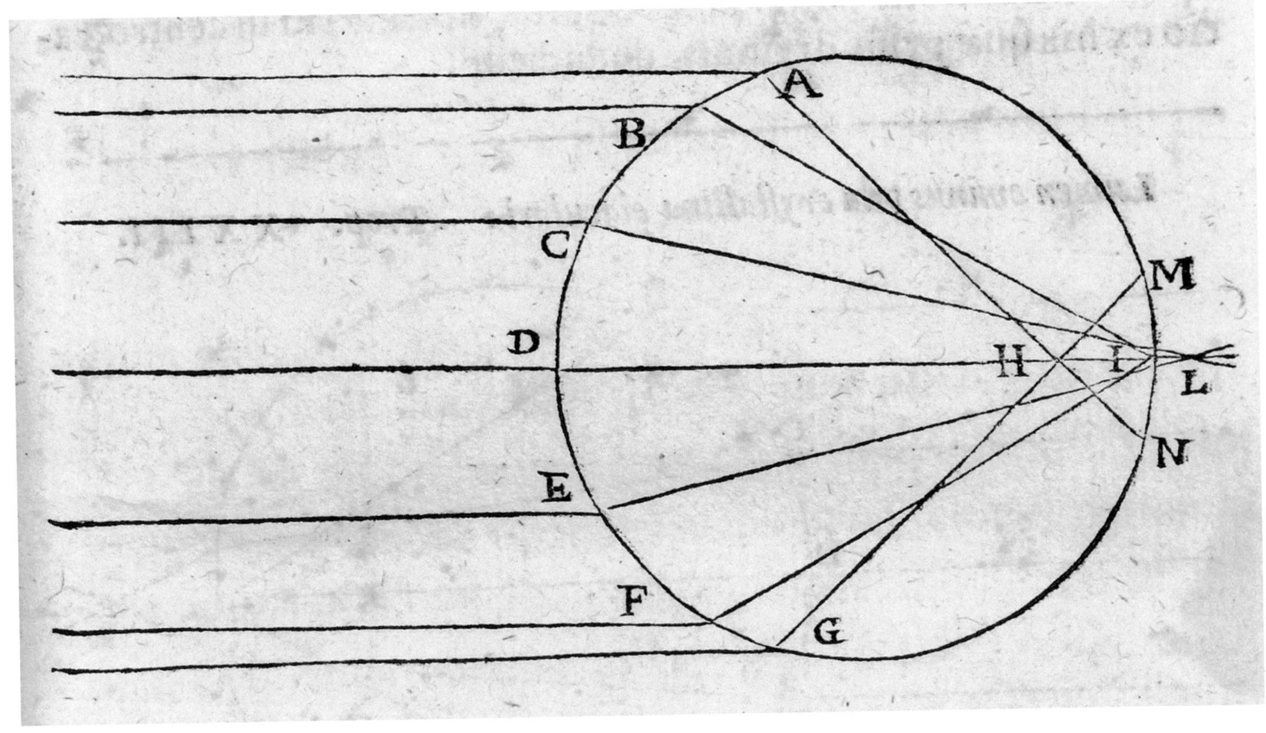
A
B
C
D
E
F
G
H
I
L
M
N

Isaac Newton

Opticks

London, England, 1718

Sir Isaac Newton (1642–1726) deserves his reputation as one of the most brilliant and extraordinary scientific minds of all time. He made significant contributions to physics, mathematics, astronomy, and chemistry, though his less-known interests in and attention to alchemy and theology occupied as much of his time and more of his writings. His religious views, including his rejection of the Trinity, were unorthodox if not heretical.

At Cambridge University, Newton lectured on optics between 1670 and 1672. From his experiments on the refraction of light by prisms, he inferred (correctly) that white light consists of light of many colors, which can also be recombined into white light. He further concluded (incorrectly) that the chromatic aberration of light so dispersed could not be overcome. In response, he developed a reflecting telescope that gathers light with a mirror instead of a lens.

The notes from his investigations were eventually compiled and published as *Opticks* (1704), of which a second edition appeared in 1718, a third in 1721, and a fourth in 1730. Each edition concluded with a set of questions or queries that evolved from short discussions into complex essays by the fourth edition. Newton promoted the idea that light consists of particles (or corpuscles, as he called them) obeying the laws of mechanics that he had developed. He also put forward the notion of light as waves to explain diffraction.

This image, from the second edition, shows the separation of white light into a spectrum of its component colors as a result of a curved lens placed on a surface. Though first described by Robert Hooke in 1664, this phenomenon is known as Newton's rings and is the result of the constructive and destructive interference of the waves of light. Similar interference patterns appear on soap bubbles or oil drops.

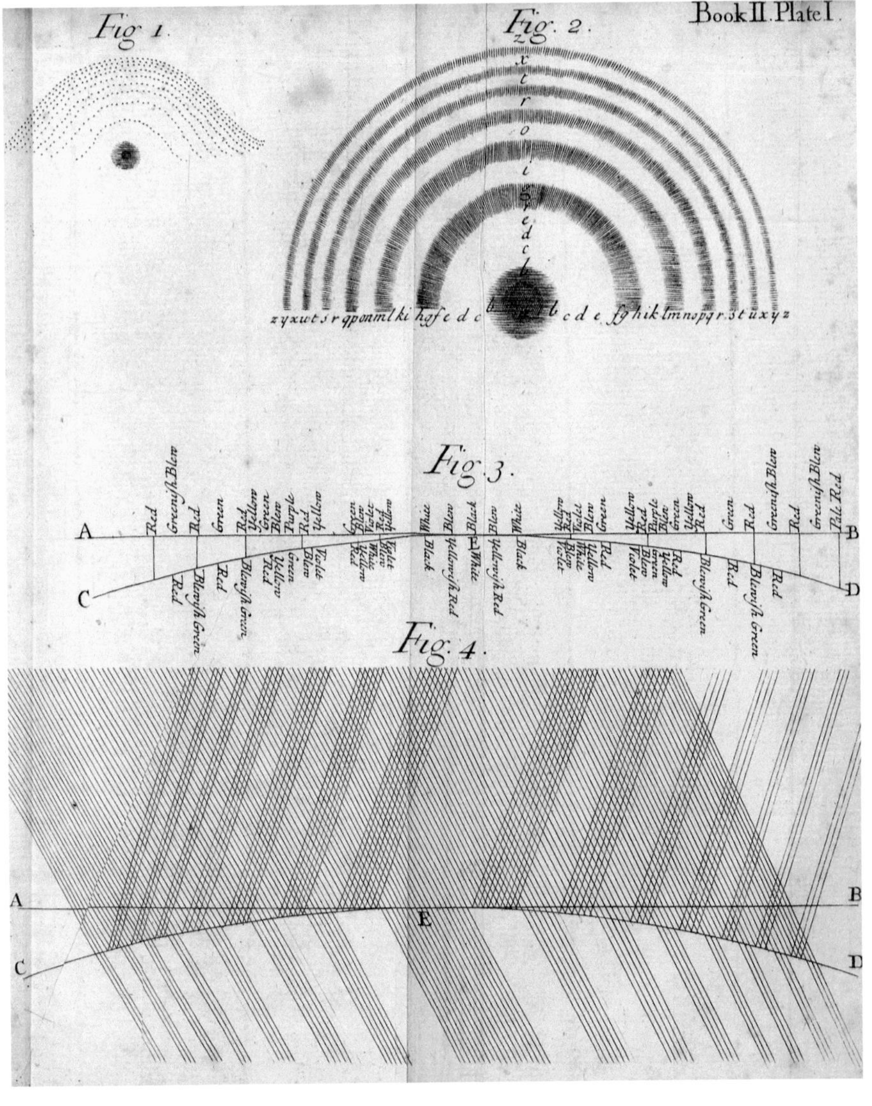
Book II. Plate I.
Fig. 1.
Fig. 2.
Fig. 3.
Fig. 4.
A
B
C
D
E

James Gregory *Optica Promota* **London, England, 1663 QC353 .G74 1663**

The problems of spherical and chromatic aberrations, and the imperfect images they produced, led astronomers of the 17th century to explore possible techniques for overcoming the limitations of lens-based refracting telescopes. When Isaac Newton pronounced the impossibility of any solution, he proposed using a mirror-based reflecting telescope instead and produced an example in 1668. Italian Jesuit Niccolò Zucchi (1586–1670) had built a reflector earlier, but it was not a high-quality instrument.

Also prior to Newton, James Gregory (1638–1675) had proposed the same solution and published the earliest illustration (shown) of a reflecting telescope in his first published work, *Optica Promota*. Gregory, a Scottish mathematician, taught at St. Andrews and Edinburgh Universities, publishing several important works in mathematics before dying tragically at the age of 37.

Most reflecting telescopes use a primary mirror to gather light and a secondary mirror that brings the light to an eyepiece, usually consisting of a lens or lens system. Gregory's design precedes two other reflecting formats, the Newtonian design, which uses a flat secondary mirror and an eyepiece at the top of the telescope, and the Cassegrain design, which is similar to Gregory's but uses a convex hyperbolic mirror for the secondary rather than Gregory's concave spherical.

Due to the difficulty in shaping the mirrors for such a telescope, the first Gregorian reflector was not built until 1674, but it became the format for James Short and other makers as the standard reflecting telescope until the 19th century. Today's research telescopes are invariably reflectors and use techniques extending Gregory's basic design.

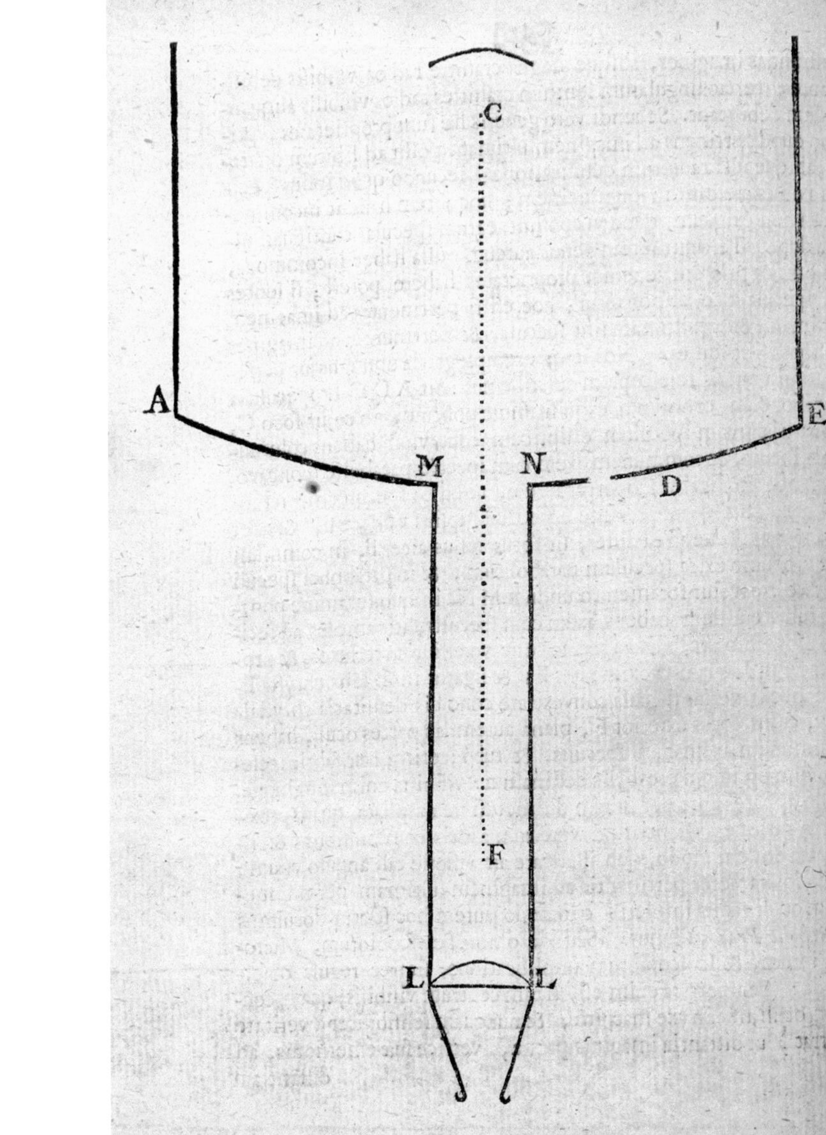
C
A
E
M
N
D
F
L
L

Ephraim Chambers

"Optics Tab. VI" from *Cyclopaedia: Or, A Universal Dictionary of Arts and Sciences*

London, England, c. 1728–1786 P-8s

Although encyclopedias were not uncommon in western Europe before the seventeenth century, the first encyclopedia of science appeared due to the efforts of Ephraim Chambers (1680–1740) under the title *Cyclopaedia: Or, A Universal Dictionary of Arts and Sciences*. At least five editions appeared between 1728 and 1743. Its translation and popularity inspired the encyclopedic movement of the eighteenth century, and in particular the *Encyclopédie* by Denis Diderot and Jean le Rond d'Alembert, which began as a translation of Chambers's efforts.

Chambers's work included many technical details and a wealth of illustrations. This image shows the fundamental features of optical properties of the eye, of single, double, and triple lenses, and of several telescopes, including Huygens's aerial refracting telescope and reflectors.

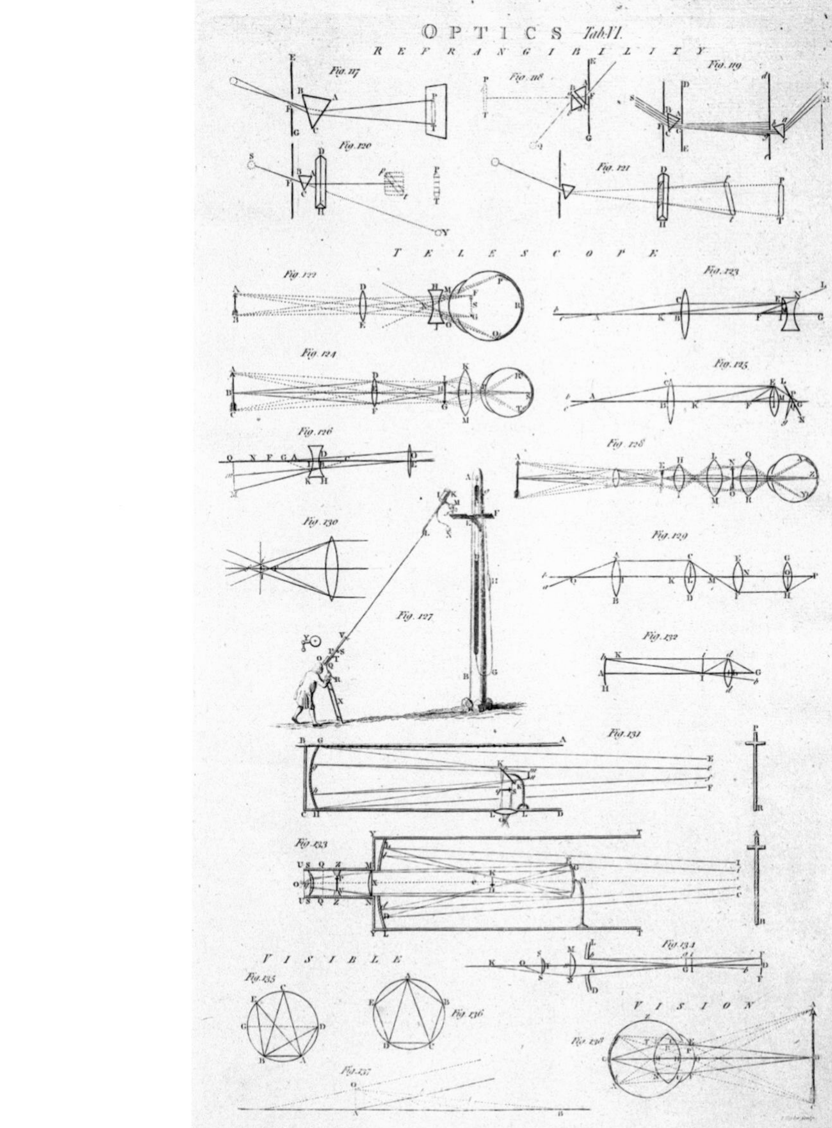

OPTICS Tab.VI.
REFRANGIBILITY
TELESCOPE
VISIBLE
VISION

George Adams

Reflecting Telescope with Tripod

London, England, c. 1748–1755 A-352

This late-18th-century brass reflecting telescope, with stand and wooden case, shows many of the interesting features typical of its time. The original box survives, along with a description of its use on the underside of the lid. The sketch on the left of the box lid shows how it operates; light comes in from the right, goes down the length of the whole tube to the primary mirror, bounces back to the right off a much smaller secondary mirror, and then out through the eyepiece at the far left. All Gregorian style reflectors use this light path. The bottom of the brass telescope stand ends in a screw, which can be used to attach such a telescope to a wooden fence post or to a tree trunk.

Documentation of the telescope included on the inside of the telescope's case shows repairs made in 1770 and later repairs to the decorative barrel paint. The interesting swirly patterns on the main part of the tube show a change in fashion from the bright shiny brass of earlier days to a sort of fake marbled paper look preferred around 1800, when this telescope was presumably altered.

The outside of the primary mirror, at the very left, includes a signature by the maker, George Adams (d. 1773) of London. Street addresses at that time did not include numbers, so Adams used "the head of Tycho Brahe" to indicate his shop.

Claude-Siméon Passement

Description et usage des telescopes

Amsterdam, the Netherlands, 1741

Claude-Siméon Passement (1702–1769) wrote books on telescopes (and microscopes) that went into several French and German editions. In the 18th century, someone who purchased a scientific instrument often bought a descriptive manual on how it was made and used. Such a book might also feature a lengthy description of what one could observe with the device. For those unable to afford the much more expensive instrument itself, the text provided an opportunity for learning and enrichment.

Passement's *Description et usage des telescopes* includes this detailed foldout illustration depicting one of his telescopes and its component parts. At the top, we see the thick metal disk of the speculum mirror (Figure 4), while the middle of the page shows the eyepiece unit that screws into the speculum. This example is an early one, as indicated by the absence of any device to help aim the telescope towards the desired object. Later reflectors include a "finder" that consists of aligned holes or even a small telescope for locating celestial targets. Just below the assembled telescope on the right appears the mechanical arrangement of the secondary mirror, which slides up and down the tube to allow for focusing. This Gregorian style is used for the overwhelming majority of telescopes sold in the 18th century.

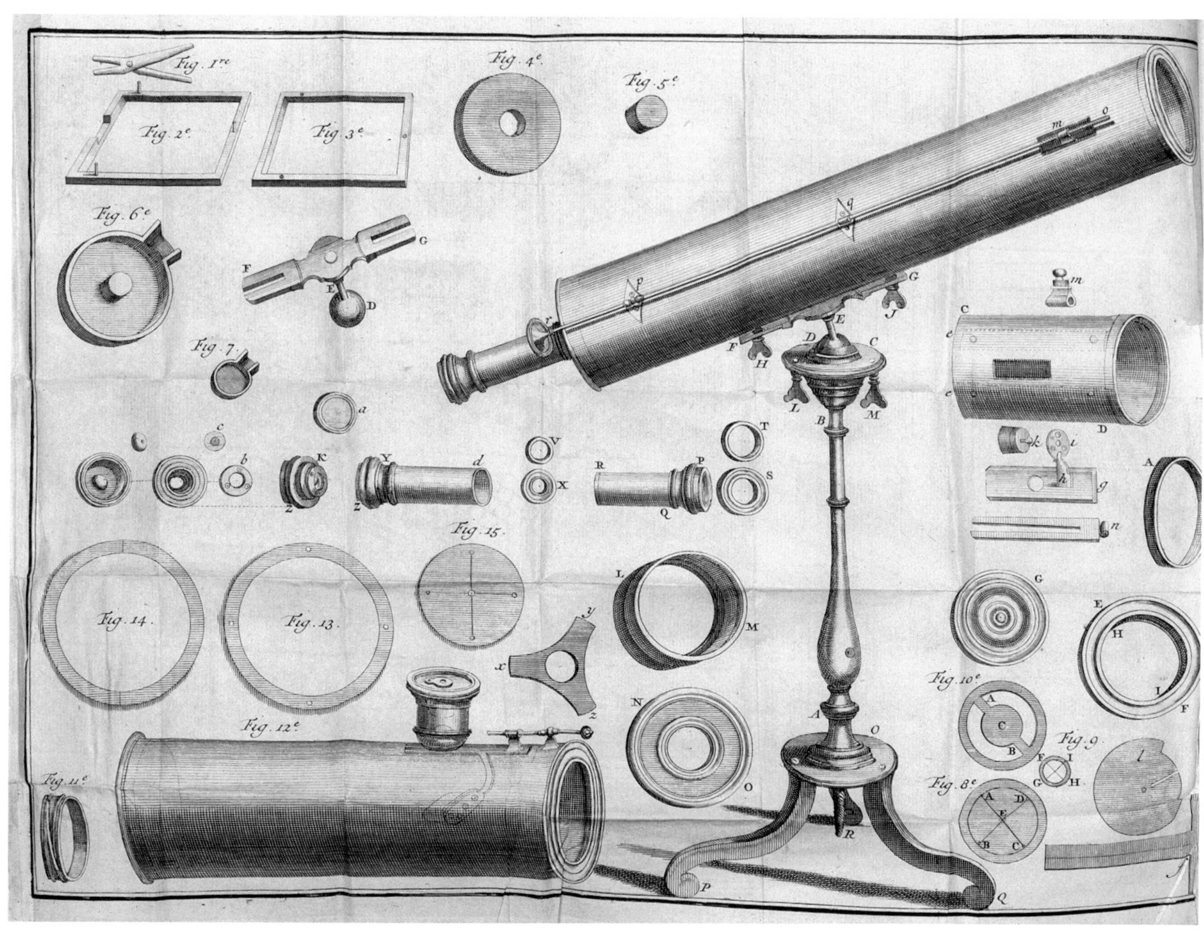
Fig. 1re
Fig. 2e
Fig. 3e
Fig. 4e
Fig. 5e
Fig. 6e
Fig. 7
Fig. 8e
Fig. 9
Fig. 10e
Fig. 11e
Fig. 12e
Fig. 13
Fig. 14
Fig. 15

James Short

Reflecting Telescope

London, England, 1756 W-32

James Short (1710–1768), a Scottish mathematician and optician, became the leading maker of reflecting telescopes. Born in Edinburgh, he began his telescope-making operations there before moving to London. His earliest telescopes used glass mirrors, as recommended by fellow Scotsman James Gregory, who designed the reflecting telescope style named after him. Virtually all of Short's telescopes use the Gregorian design, but the telescopes he made professionally used metal mirrors. Short developed practical techniques that made his mirrors by far the best of his day, enabling him to become wealthy from the sale of his telescopes.

Like his others, this Short telescope features his unique signature "formula" on the end of the barrel near the eyepiece: "James Short London 124/904=7". This means that this was the 904th telescope he had made, and the 124th with a focal length of 7 inches. Using information about Short's production, we can date this telescope to 1756. Any serious telescope collection should have at least one instrument made by Short; this example survives with its original box.

* JAMES SHORT LONDON 124/904 = 7.

Denis Diderot and Jean le Rond d'Alembert

"Optique" from *Encyclopédie, ou dictionnaire raisonné des sciences, des arts, et des métiers*

Paris, France, 1777 P-72f

The *Encyclopédie (or systematic dictionary of the sciences, arts, and crafts*), a general encyclopedic dictionary published between 1751 and 1772 and edited by Denis Diderot (1713–1784) and Jean le Rond d'Alembert (1717–1783), consists of 35 volumes and almost 72,000 articles. Often considered the embodiment of the Enlightenment, it represents the efforts of mid-18th-century intellectuals to define and refine practical and theoretical knowledge, enabling collaborative efforts to systemize and organize all knowledge.

This image, appearing under the entry "Optique", includes an illustration of a telescope and a diagram illustrating how it forms an image.

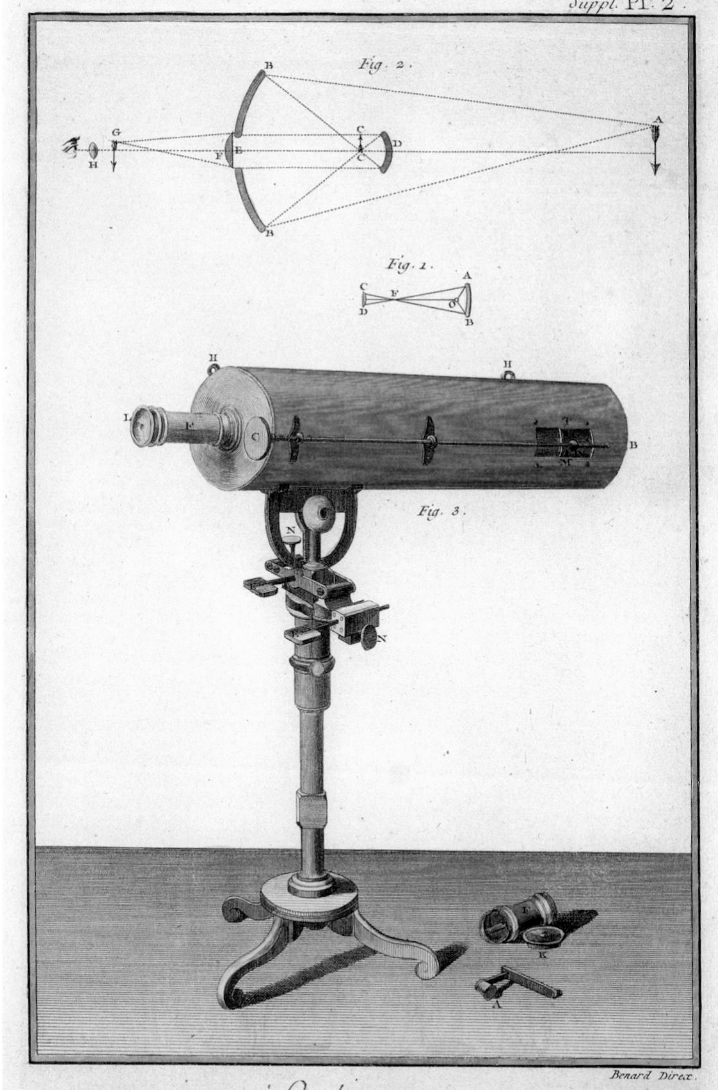
Fig. 2.
B
A
G
C
D
H
F
E
C
B
Fig. 1.
C
F
A
D
B
H
H
L
F
C
T
B
M
Fig. 3.
N
N
F
K
A
Benard Direx.

Optique.

Benjamin Martin

Frontispiece from *The Young Gentleman and Lady's Philosophy*

London, England, 1772–1781 (first published 1755) P-63a

Benjamin Martin, originally a schoolmaster, became one of the most important science popularizers of the 18th century. He also made significant improvements to the design of microscopes, including building the first ones that used achromatic lenses. His scientific instrument shop on London's Fleet Street also offered a variety of large and pocket-sized microscopes, telescopes, planetarium machines, and other scientific instruments, as well as his own popular books.

From 1755 to 1763, Martin also devoted much time to the *General Magazine*. The main section of this monthly project featured a fictional story called "The Young Gentleman and Lady's Philosophy." In each issue, university student Cleonicus instructed his sister Euphrosine on astronomy, globes, optics, telescopes, and other scientific subjects. In the mid-18th century, women like Euphrosine could not attend universities, so she was grateful for her brother's lessons, commenting on "how happy will be the age when the ladies may modestly pretend to knowledge." Here, Cleonicus and Euphrosine spend an evening session in their library, a scene likely repeated in numerous wealthy households to gain the rudimentary astronomical knowledge expected in proper social circles.

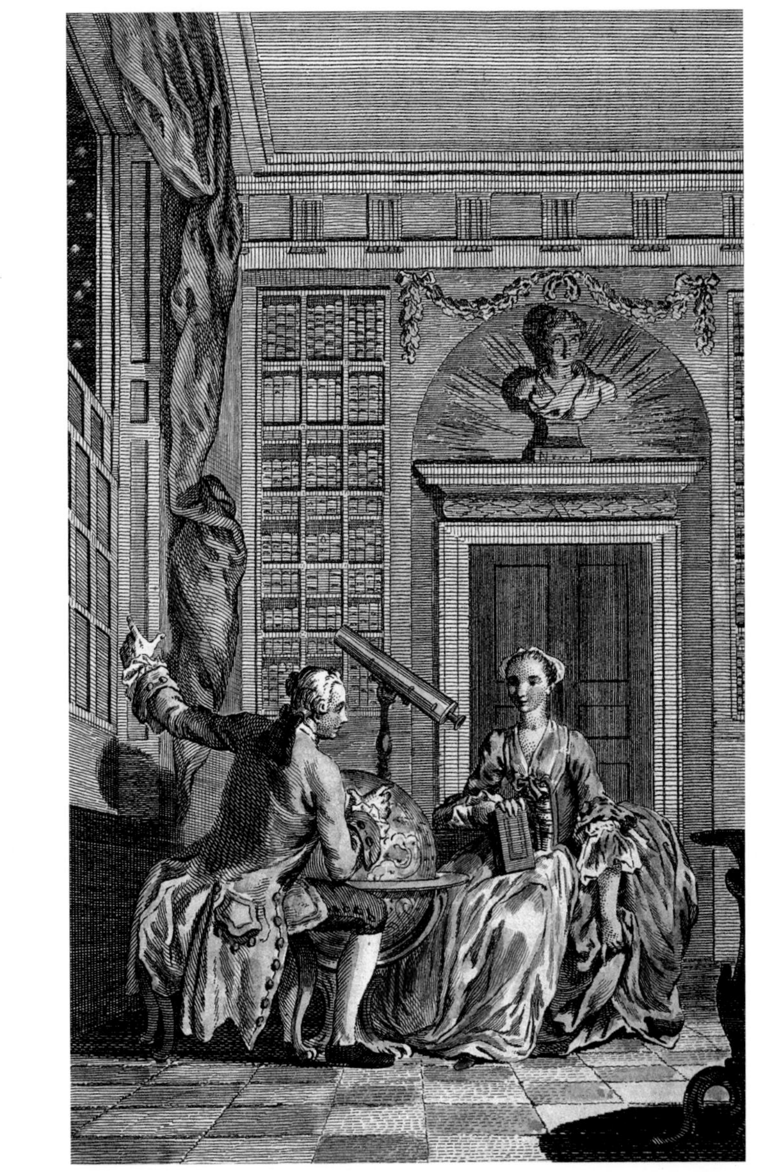

James Ferguson

Lectures on Select Subjects

London, England, 1772

James Ferguson (1710–1776) first learned about the stars while tending sheep as a 10-year-old boy. He taught himself to read and filled his empty hours making models of clocks, mills, and other machines. In 1734, he moved to Edinburgh, where he supported his family by drawing miniature portraits. After spending a few years in Inverness, he moved in 1743 to London, where he resided the rest of his life. In 1748, he began to give public lectures on science, especially astronomy. His clear lectures, supplemented with several printed editions of those lectures, complete with numerous diagrams and illustrations, drew large audiences around England and made him the most recognized science popularizer of the day.

Ferguson also invented and improved a variety of scientific instruments, astronomical devices in particular. This image shows various optical setups, including singlet and doublet lenses, as well as the optics of the eye. So clear and so thorough were his printed lectures that William Herschel, the most important astronomer of the late 18th century, commented that he learned his astronomy from Ferguson, one of his favorite authors.

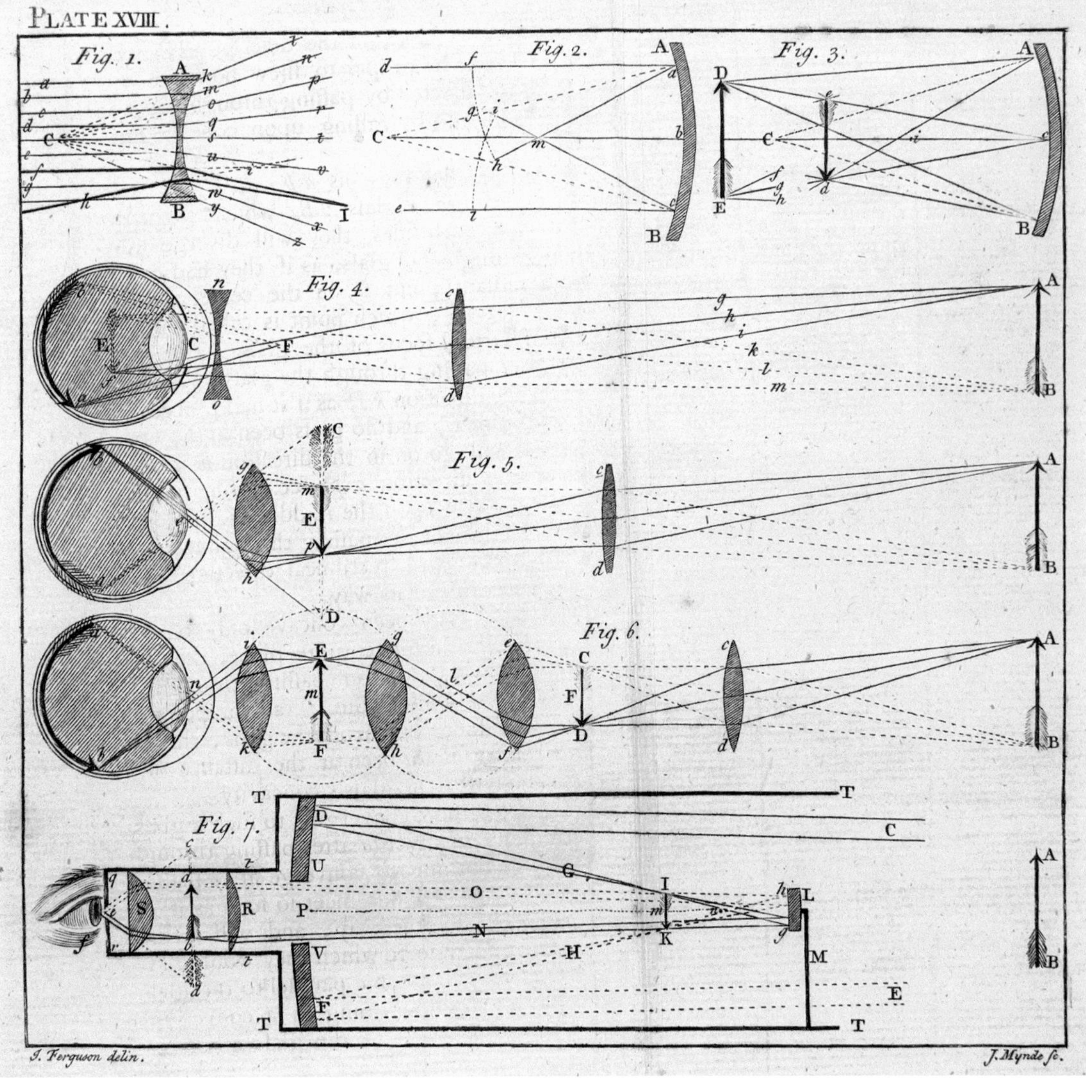
PLATE XVIII.
Fig. 1.
Fig. 2.
Fig. 3.
Fig. 4.
Fig. 5.
Fig. 6.
Fig. 7.
J. Ferguson delin.
J. Mynde sc.

Unidentified Maker **Refracting Telescope** **France, c. 1750 M-433**

From its invention in 1608, the telescope was immediately recognized as a tool for astronomy, military surveillance, education, and entertainment. Most frequently associated with astronomy, it came to symbolize discovery and knowledge. Most telescopes for popular audiences, whether fitted for sturdy stands or sized to fit in pockets, offered little hope of seeing much.

Pocket telescopes served as wonderful novelties that launched many polite conversations. This example, most likely French but possibly English, survives with its original red-satin-lined, hinged, flip-top case to protect it from the other objects in an 18th-century gentleman's pocket. The main tube is covered in black shagreen leather, with a decorative band of small brass studs at both ends. The objective glass is mounted in mahogany, with a gilt brass, screw-on rim, worked in decorative floral motif. The tube is pasteboard, covered in green vellum, painted black inside; its tooling resembles M-442 (see No. 46). The image is clear, but is of little use due to its very small size and magnification.

Even so, its unveiling at a social event, such as opera, would draw positive attention to its owner. Numerous styles of small telescopes, including so-called opera glasses, evolved over the years, as did the opera, which first appeared at the same time telescopes did. The earliest advertisement for an opera glass dates from 1730, but the binocular form did not appear until around 1825.

William Harris & Co. **Pocket Refracting Telescope** **London, England, c. 1830 N-60**

Prior to his career making scientific instruments as an independent retailer, William Harris worked for optical researcher Sir David Brewster in Edinburgh until about 1800. Harris produced a variety of mathematical and drawing instruments, as well as telescopes, and around 1814, he opened a shop at 50 Holborn in London. His earlier products are marked “William Harris and Co.”, which changed around 1840 to reflect the firm's renaming as “William Harris & Son.” This establishment is listed among instrument makers with products on display at the Great Exhibition of 1851.

Many of the telescopes produced by the Harris firm were optical toys, such as pocket telescopes and walking-stick telescopes. This small telescope, possibly used as an opera glass, consists of silver draws, the main one covered with a dark horn collar, split by shrinkage. The achromatic objective lens is silver mounted. The eyepiece end features the maker's inscription: “WM. Harris & Co. 50 Holborn. London”. This last unit unscrews, revealing the brass-mounted-eyepiece lens, held in place by a screw-in brass rim. The six silver draw tubes all have internal stops, eliminating overextension. The fitted wooden case is covered in red tooled leather. It is hinged on one side to open like a clamshell, with a push-button catch on the opposite side, and lined with purple velvet and white satin. While not a powerful instrument, its elegance and sophistication make it a suitable accessory for any gentleman.

Unidentified Maker

Scene of two ladies looking through a telescope

Europe, date unknown P-172

The 18th and 19th centuries offered few opportunities for women to "improve themselves" through education. Here, two young women observe with a telescope, illustrating the impact of the telescope on all facets of society; women could be privately instructed in its use, and might employ it for their leisure. For conversations in polite company, such knowledge and experience might prove of great value.

J. Grandville

"Astronomie des dames" from *Les Etoiles*

Paris, France, c. 1845 P-266

The pseudonymous J. Grandville here depicts a salon of fashionably dressed ladies as they study the Universe with the help of a map of the Moon, an armillary sphere, diagrams, and a telescope. The "astronomy of the ladies" here involves no males, other than the *putti* in the upper right. The women lack distinction, portrayed almost identically to each other. Sketches of simple astronomical concepts appear above the ladies, while a young girl in front points to "the stars."

F. 1
F. 2
F. 3
F. 4
A
B
C
Nord
F. 5
Est
Ouest
Sud
F
F. 6
Lune horizontale
E
Lune horizontale
D
A
B
Terre
F. 7
Terre
Lune
Soleil
F. 8
Soleil
LES
ETOILE

William Nelson

Refracting Telescope

Dublin, Ireland, c. 1850 A-56

With the rise of a middle class in 18th- and 19th-century England and Ireland, merchants and gentlemen found it increasingly fashionable to be seen having a scientific object in their pocket or in their hand. The object might be a sundial, globe, watch, or even a little planetarium, but a telescope in particular was sure to draw attention. This one especially would do so. It is a fun and functional item for the scientific gentleman, who along with his social circle recognized it as a fashion accessory rather than a serious device.

William Nelson (fl. 1830–1862) was an optician, spectacle maker, and jeweler who later passed on his trade to his son in Dublin. He made this lovely refracting telescope, which doubles as a walking stick, with five sections. When the upper two sections are unscrewed from the bottom three sections, the brass eyepiece of the telescope is revealed. The knob at the top of the telescopic walking stick unscrews to expose an achromatic objective lens. The image it produces is not especially good, but that purpose was not the primary goal of making, owning, or showing off this delightful object.

J. Grandville

Les Métamorphoses du jour

Paris, France, 1828–1829 P-9

Widely known as J. Grandville, French caricaturist Jean Ignace Isidore Gérard (1803–1847) first earned widespread recognition for his *Les Métamorphoses du jour* (1828–1829). This comedy consists of a series of 70 scenes employing animal features, particularly faces, to portray human characteristics and foibles. His success led to the publication of numerous illustrations, especially political caricatures and satirical sketches, in such periodicals as *Le Silhouette*, *L'Artiste*, *La Caricature*, and *Le Charivari*. The censorship of caricatures in 1835 led Gérard to continue his career as a book illustrator.

This image, number 57 in the series *Les Métamorphoses du jour*, shows four individuals, two observing the night sky, one of them explaining that by looking through the telescope the other can observe the constellation Capricorn the Goat. The wisdom of the bird and earnest attention of the hunched-over rhinoceros depict and exaggerate the pretentious attitudes of newly middle-class 19th-century social climbers who cannot, in the end, escape their humble origins. The caption at the bottom reads: "A votre droite est le signe du Capricorne" – "To your right is the constellation Capricorn", to which the gentleman pays rapt attention while evidently unaware of his companion's own interest elsewhere.

Les Métamorphoses du Jour.
N.° 57
J. Grandville.
Lith de Langlumé
A votre droite est le signe du Capricorne.
chez Bulla rue S.t Jacques N.° 38.
et chez Martinet rue du Coq.

Angelo Deregni

Reflecting Telescope

Italy, c. 1780 M-449

Along with Semitecolo, Olivo, and da Silva, Angelo Deregni (fl. 1780, d. 1820) was one of several makers of cheap, popular telescopes in northern Italy in the late-18th and early-19th centuries. These mass-produced items flooded the market, with many examples surviving in museum and private collections today. Even after the introduction of the achromatic doublet lens in the 1750s, Venetian telescopes continued to provide the cheaper and "old-fashioned" singlet objectives.

Typical of the cheap Venetian telescopes, this one consists of pasteboard tubes covered with vellum, and ebony or horn lens mounts. The lenses are original, with one possible exception in the eyepiece. Each end is protected with a brass cap, with dust sliders. The main tube is decorated with circumscribed blind tooling and scrolling foliage patterns. The draws are partially covered with white vellum and have simple gold-tooled lines. The signature appears on the main tube.

M449

Unidentified Maker **Reflecting Telescope** **Nuremberg, Germany c. 1790 W-310**

From the turn of the first millennium, the location of Nuremberg on major trade routes from Italy to northern Europe enabled it to grow in size and in wealth, becoming the center of the German Renaissance in the 15th and 16th centuries. Its cultural flowering in the sciences and humanities also extended to its extensive printing industry. Astronomy was also well represented. Regiomontanus built an important observatory here at the end of the 15th century, just prior to the publication of the first printed star charts in 1515 by Albrecht Dürer and a half century before the central portion of Copernicus's *De Revolutionibus* was printed here in 1543. Nuremberg was especially known for ivory diptych sundials as well. Toward the end of the 18th century, as several European cities developed a market for cheap telescopes, instrument makers emerged to satisfy the public demand. In addition to London and Venice, one such city was Nuremberg.

The main tube of this example is covered with black leather, with a seam that runs diagonally rather than parallel to the tube, as is usual. Each tube originally included simple brass fittings or ferrules with milling, though they are missing on the last two of the four draws. The tubes are covered not by paste paper but by a woven paper with wood grain decoration, a fashion that started around the end of the 18th century. The original lenses, each held in place by a split ring, yield a nice, clear image with magnification of 6x.

Dudley Adams

Patent for Refracting Telescope **London, England, c. 1800 MS 48**

Refracting Telescope **London, England, c. 1800 A-410**

This early-19th-century English metal refracting telescope is signed by the well-known instrument maker Dudley Adams, the son of George Adams. He was, as were his father and brother George, optical and mathematical instrument maker to England's King George III and optician to the Prince of Wales. Dudley also made a variety of other instruments as well, such as thermometers, barometers, and globes in particular.

Like other craftsmen of the day, Adams tried to improve his products with innovative techniques that would lead to new patents, improved reputation, brand recognition, and increased profits. Adams was granted a patent in 1800 for an improved method of connecting draw tubes; his invention enabled his telescopes to collapse to a very short length and yet extend in a stable manner to a length that provided a high-quality image.

This 2-foot telescope uses his patented draw tube, displaying his signature on the eyepiece tube. All of the tubes are brass, though the grip of the main tube is wood, possibly mahogany. It survives with all of its original lenses.

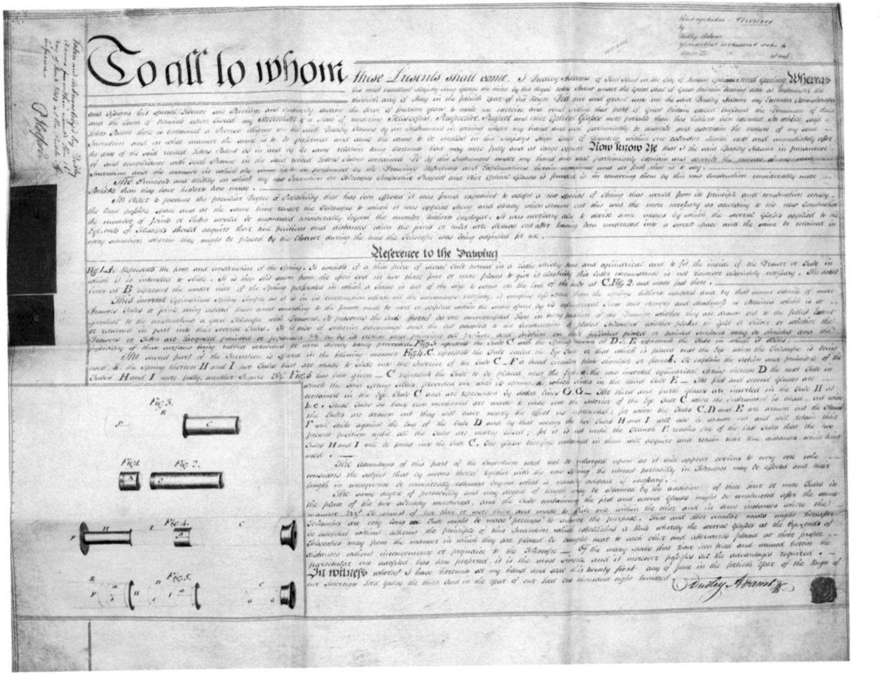

To all to whom these Presents shall come, I Dudley Adams ...

Now know Ye

Reference to the Drawing

Fig. 1. A. Fig. 2. Fig. 3. Fig. 4. Fig. 5.

In witness whereof

Dudley Adams

Unidentified Maker

Refracting Telescope

England, c. 1810 M-450

Straw marquetry, a craft similar to wood marquetry using straw rather than wood veneer, likely came from eastern Europe to England in the 17th century. Of the many English camps built for prisoners captured in the Revolutionary and Napoleonic Wars (1793–1815), the one most famous for straw marquetry was Norman Cross (just west of Cambridge). Many prisoners spent their empty hours by making decorative ornaments from wood, bone, and straw marquetry. The regular market beside the prison gates offered these products for sale, enabling prisoners to purchase alcohol, tobacco, and other items. Many of these objects appear today in museum galleries and private collections.

The main tube of this refracting telescope is covered with geometric mosaic designs made from small straw pieces of various colors. Mosaic straw pieces were made by bundling dyed straw, then slicing the straw to get a thin cross-wise multicolored sheet. This sheet was then glued to objects, such as sewing boxes and other containers. Some of these containers were to protect eyeglasses and telescopes, but this example is quite rare because the object itself has been decorated in this fashion. The telescope has four brick red draw tubes that are also partially covered in straw. The objective lens and the eyepiece lenses are missing, but the original horn mounts remain intact. The protective end caps also survive.

W. & S. Jones

A catalogue of optical, mathematical, and philosophical instruments

London, England, 1808

The partnership of William Jones (1763–1831) and his younger brother Samuel lasted for several decades. Opticians and instrument makers themselves, their firm, W. & S. Jones, was one of the most respected and successful among the London scientific instrument makers. They were among the first to combine the production and sale of instruments, which included microscopes, telescopes, planetariums, and globes, as well as much more. Like their competitors, they also affixed their name to products made by other makers.

This catalogue of their wares provides a glimpse into the life of the retail world of 200 years ago. The firm offered an astonishing variety of spectacles, refractors, reflectors, and microscopes for diverse budgets and with an array of ornamentation. Spectacles could cost a shilling or two, making them widely available. Simple telescopes and microscopes started at about a pound (20 shillings or 240 pennies), with good ones going for much more, pricing them beyond the average laborer but well within reach of the middle-class merchant or gentleman.

A

CATALOGUE

OF

Optical, Mathematical, and Philosophical Instruments,

MADE AND SOLD BY

W. AND S. JONES,

[No. 30,]

LOWER HOLBORN, LONDON,

(Removed from their old Shop, No. 135, next to Furnival's Inn.)

OPTICAL INSTRUMENTS.	£.	s.	d.
BEST double-jointed standard gold spectacles, with pebbles, and fish-skin gold-mounted case.	16	16	0
Ditto, single-jointed, with ditto case.	10	10	0
Best double-jointed silver ditto, with pebbles	1	16	0
Ditto, ditto, with glasses	1	1	0
Best single-jointed, with pebbles	1	8	0
Ditto, with glasses	0	13	0
Best double-jointed steel ditto, with glasses	0	9	0
Second best double-jointed steel spectacles, with spring case	0	7	6
Common ditto	0	4	6
Best single-jointed steel spectacles, with fish-skin case	0	5	6
Second best ditto	0	2	6
Common ditto	0	1	6
Tortoiseshell spectacles, silver-jointed, with pointed and other shaped sides, peculiar for their lightness and uninterruption of dressed hair, in morocco leather cases.	0	10	6
Ditto, double-jointed frames	0	15	0
Spectacles for eyes that have been couched	0	7	6
Ditto, with green glasses for very weak and inflamed eyes, according to the frames, from 6*s.* to.	1	1	0
Ditto, for the same purpose, with new contrived portable shades to screen the eyes from candle, or other light.	0	15	0
Nose spectacles in silver	0	7	6
Ditto, in tortoiseshell and silver	0	4	6
Spectacle cases in great variety, from 2*d.* each to.	3	3	0
Concave glasses for short-sighted persons, in horn cases.	0	1	6
Ditto, in tortoiseshell, pearl, silver, &c. from 2*s.* 6*d.* to	2	2	0
Ditto in new contrived frames for shooting caps.	0	16	0
Reading and burning glasses in various mountings, from 3*s.* to	1	16	0
Convex glasses for watch-makers, engravers, &c. from 1 .to	0	10	6
Gogglers, to guard the eyes from the dust or wind.	0	3	6
New green light shades for the eyes.	0	6	6

Printed by W. Glendinning, 25, Hatton Garden, 1808

Johann Michael Voltz **Craftsmen in a workshop** **Germany, c. 1820 P-185**

Bavarian artist Johann Michael Voltz (1784–1858) drew diverse scenes of nature and science, and he opposed the Napoleonic war machine that ravaged the countryside. He is well-known for his depictions of the Russian army, c. 1805, for a series of plates for the "Augsburg Uniform Series," and for domestic scenes as well. His illustrations have earned a reputation for their accurate portrayal of even the most minute details.

This sketch brings the viewer into an early-19th-century scientific shop. On the left, drawing tools adorn the wall, with a terrestrial globe conveying the diverse skills of the shop's craftsmen. A reflecting telescope on the ground at right, and a large example on a stand in the center, display their optical expertise, as does the careful attention being paid to the half-circle angle-measuring device needed to make precise astronomical observations.

inv. Voltz.

Unidentified Artist

"Street Telescope Exhibitor" from Henry Mayhew, *London Labour and the London Poor*

London, England, 1862 P-231

In 1862, Henry Mayhew published an extraordinary survey of *London Labour and the London Poor.* In the first volume, he provides a thorough description of costermongers or patterers, those people who make their living on the streets of London by offering a wide variety of products and services. Although Mayhew divided this population into six classes, they fall into three overarching categories: those who sell products, those who exhibit a show, or those who beg.

In the second group fall jugglers, clowns, showmen, fortune-tellers, and those who invite the public to enjoy the wonders unveiled by the telescope or microscope. Special astronomical events always attract an audience, but a large telescope set up on a street at any time was sure to draw everyone's attention. In the 19th century, an entrepreneur might make a reasonable income from putting out a telescope on the street and charging passersby a modest fee for a look at the heavens. Here, one such enterprising owner of a very large telescope charges for a "telescope view of the Moon".

STREET TELESCOPE EXHIBITOR.

[From a Photograph.]

Thomas Wright

An Original Theory or a New Hypothesis of the Universe

London, England, 1750 **QB42 .W7 1750**

Thomas Wright (1711–1786) is best known as an astronomer, but made his living as a teacher. He also designed English gardens and worked as an architect, which provided him with expertise that yielded his most important astronomical insight, which appeared in this volume. In it, he suggests that the appearance of the Milky Way is "an optical effect due to our immersion in what locally approximates to a flat layer of stars." In other words, if we are located somewhere in a thin, flat layer of stars, we would see a lot of stars when we look in the direction of that layer, but very few stars if we look out of the layer. That thin layer might be a plank, as in this sketch, or it might be part of a large sphere of stars, like the skin of a large ball, as in other sketches in this volume.

Either way, his suggestion was noted by philosopher Immanuel Kant and later by astronomer William Herschel, who made it the foundation of his astronomical observing program. Herschel spent years observing the sky in all directions to see if he could sketch out the details of the Milky Way and came up with two distinct models. Wright's other notable suggestion was that faint patches of light known as nebulae might actually be distant galaxies similar to our own. It would take more than a century and a half to determine that Wright's insight was fundamentally correct.

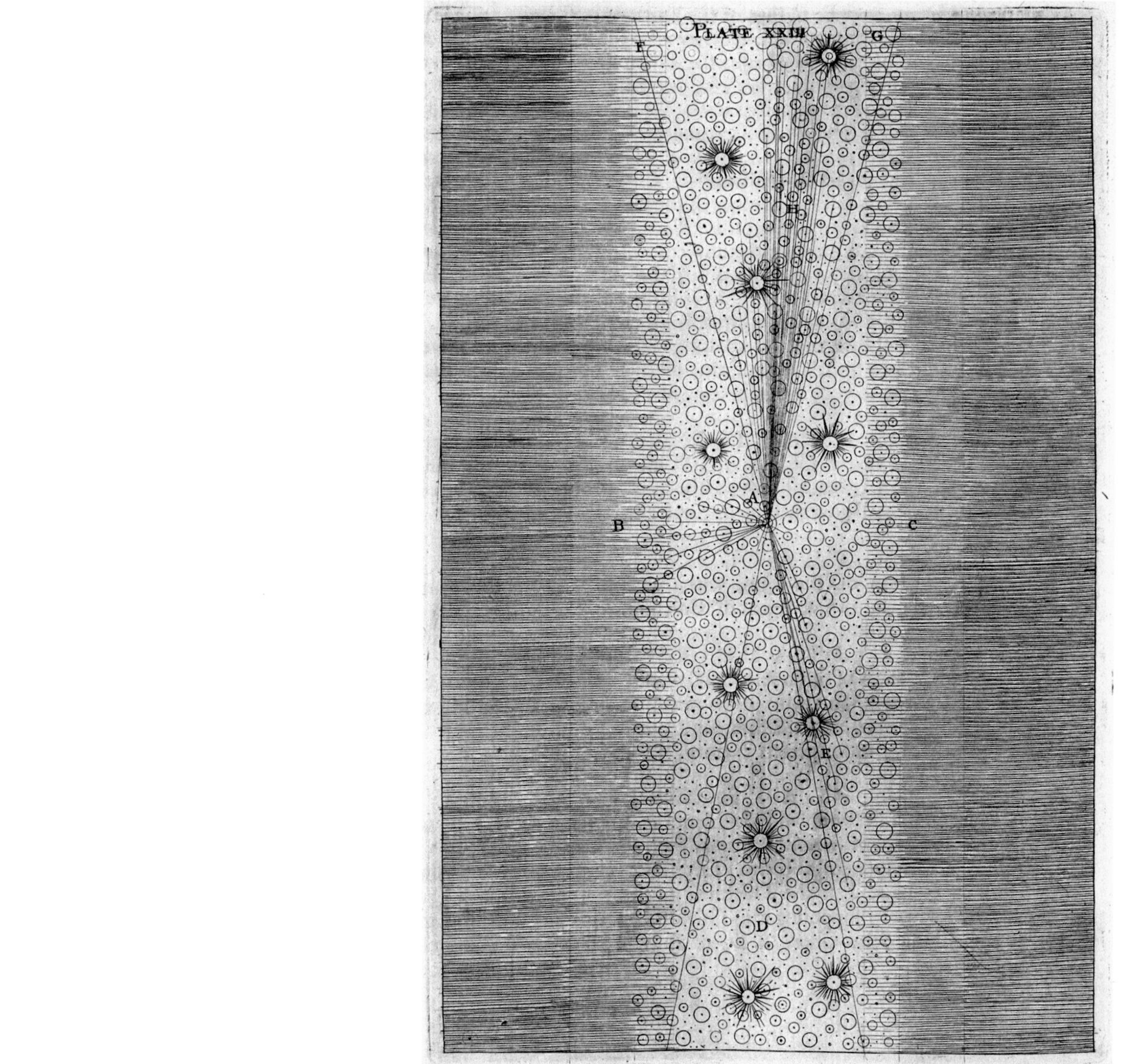
PLATE XXIII
F
G
H
A
B
C
E
D

Frederick Rehberg

Portrait of "Dr. Herschel"

London, England, 1814 P-221

Frederick Rehberg (1758–1835) is noted for his portraits of famous and lesser-known people, as well as for his portrayals of nature. By 1814, William Herschel (1738–1822) was one of the most recognized men of science in Europe. Rehberg portrayed Herschel as the 76-year-old astronomer he had become. In 1781, Herschel accidentally discovered Uranus when it was located in the constellation Gemini, visible here as the background to his portrait.

Johann Elert Bode

Allgemeine Beschreibung und Nachweisung der Gestirne

Berlin, Germany, 1801 QB6 .B63 1801

Johann Elert Bode (1747–1826), a German astronomer, published important celestial atlases. He is most commonly recognized for his expression of the Titius-Bode law, which provides a simple mathematical formula for describing the distances from the Sun to the classical planets. His atlases include data from Flamsteed's atlas and others as well. His 1801 *Uranographia* was accompanied by the illustrated volume (*General Description and Report of the Stars*), which listed the coordinates over 17,000 stars.

Bode sent this copy directly to William Herschel, who used it as a reference manual for more than 20 years. To make it more usable, Herschel (or more probably his sister Caroline) has pasted in a conversion table for switching between two coordinate systems (from degrees/minutes/seconds to house/minutes/seconds). He (or she) has also provided an index to the book, enabling the user to locate the tables for the constellations.

Throughout the volume, Herschel has made additions and edits, in some places noting that a listed star is actually a double star, likely visible only with his telescope. Other places feature corrections and other comments. The most pointed remark is on this title page: "Stupid!" This refers to the coordinates having been adjusted (reduced) to the common date 1801 rather than 1800.

ALLGEMEINE

BESCHREIBUNG UND NACHWEISUNG

DER

GESTIRNE

NEBST

VERZEICHNISS

DER

GERADEN AUFSTEIGUNG UND ABWEICHUNG VON 17240 STERNEN, DOPPELSTERNEN, NEBELFLECKEN UND STERNHAUFEN.

VON

J. E. BODE,

KÖNIGL. ASTRONOM, MITGLIED DER AKADEMIEN UND SOCIETÄTEN DER WISSENSCHAFTEN ZU BERLIN, LONDON, PETERSBURG, STOCKHOLM UND UTRECHT, WIE AUCH DER BERLINISCHEN GESELLSCHAFT NATURFORSCHENDER FREUNDE.

(ZU DESSEN URANOGRAPHIE GEHÖRIG.)

BERLIN 1801.

BEYM VERFASSER.

Received Jan 1. 1801. Stupid!

DESCRIPTION

ET

CONNOISSANCE GÉNÉRALE

DES

CONSTELLATIONS

AVEC

UN CATALOGUE DE L'ASCENSION DROITE ET DE LA DÉCLINAISON DE 17240 ÉTOILES, DOUBLES, NÉBULEUSES ET AMAS D'ÉTOILES.

PAR

J. E. BODE,

ASTRONOME ROYAL, MEMBRE DES ACADÉMIES ET SOCIÉTÉS DES SCIENCES DE BERLIN, LONDRES, ST. PÉTERSBOURG, STOCKHOLM ET UTRECHT, ET DE LA SOCIÉTÉ DES SCRUTATEURS DE LA NATURE A BERLIN.

J. F. Wm. Herschel

(POUR SERVIR DE SUITE A SON URANOGRAPHIE.)

Carolina Herschel

A BERLIN,
CHEZ L'AUTEUR.
1801.

To my very dear sister Isabella in great gratitude for her many years' devoted labours in the Library
W. J. Herschel
1889

Tabula IV.

VII. Caſſiopeja. Caſſiopée.

Nº.	No. F.		Magnitud.	Ascenſio recta in Temp. H.	M.	in Grad. G.	M.	S.	Var. ann. Sec.	Declinat. Boreal. G.	M.	S.	Var. ann. Sec.	Obſerv.
1	1	ε	6	XXII.	58	344	33	29	37,3	58	20	44	+19,3	L.
2	2	ð. 2½′. VI, 55. Ceph.	7	XXIII.	1	345	15	52	37,7	58	15	19	+19,4	L.
3	3	ð. 2½′. VI, 26	6	—	15	348	50	16	39,8	57	34	47	+19,7	F
4	4	d. ð. VI, 24	5	—	16	349	0	46	39,0	61	11	30	+19,7	L.
5			6	—	34	353	36	54	43,2	54	5	20	+19,9	L.
6			6	—	35	353	38	24	43,1	54	22	21	+19,9	L.
7			6	—	35	353	47	16	43,0	54	41	51	+19,9	L.
8	5	ν	5	—	37	354	20	50	42,8	57	32	35	+19,9	L.
9	6	306 Ceph.	6	—	39	354	48	7	42,7	61	6	35	+20,0	L.
10			6	—	39	354	52	18	43,4	57	52	1	+19,9	M
11			6	—	41	355	12	40	43,0	62	38	13	+19,9	L.
12			6	—	41	355	12	51	44,0	53	5	53	+19,9	L.
13			6	—	43	355	52	1	43,8	59	35	56	+19,9	L.
14	7	ρ	6	—	44	356	7	14	43,9	56	23	32	+20,0	L.
15			7	—	46	356	23	46	44,6	51	37	40	+20,0	L.
16			6	—	47	356	47	34	44,4	54	35	56	+20,0	L.
17		VI, 30	C	—	48	356	54	44	44,7	55	37	32	+20,0	H
18			6	—	48	357	1	30	44,4	49	19	40	+20,0	L.
19			7	—	49	357	10	18	45,1	52	16	52	+20,0	L.
20	8	ε ð. I, 5	6	—	49	357	14	40	44,6	54	38	52	+20,0	L.
21			7	—	49	357	17	25	44,7	52	27	22	+20,0	L.
22			5	—	52	357	53	6	44,7	60	6	52	+20,0	L.
23			6	—	52	358	5	23	43,8	59	35	44	+19,9	L.
24	9	ð. 53″. V, 79. Ceph.	6	—	54	358	30	57	45,2	61	10	50	+20,0	L.
25			6	—	55	358	43	7	43,3	60	12	21	+20,0	L.
26			7	—	55	358	50	40	45,5	52	3	50	+20,0	L.
27			6	—	56	358	53	4	45,5	45	29	16	+20,0	L.
28	10		6	—	56	359	2	55	45,3	63	3	40	+20,0	L.
29	11	β	3	—	59	359	39	26	45,7	58	3	7	+20,0	L.
30			7	O.	0	0	0	25	46,0	51	10	45	+20,0	L.
31			7	—	4	1	2	37	46,9	63	27	20	+20,0	L.
32			7	—	5	1	13	44	46,7	48	54	47	+20,0	L.
33			7	—	6	1	35	19	47,0	60	30	40	+20,0	L.
34			6	—	7	1	48	27	46,8	50	19	41	+20,0	L.
35			6	—	10	2	31	55	47,8	60	46	20	+20,0	L.
36			7	—	12	3	1	55	46,3	53	32	31	+20,0	L.
37			C	—	12	3	4	17	47,6	52	52	5	+20,0	H
38			6	—	14	3	24	2	47,1	50	55	0	+20,0	L.
39	12		6	—	14	3	25	39	45,1	60	43	35	+20,0	L.
40			7	—	14	3	36	6	47,6	51	56	35	+20,0	L.
41		nova 1572	α. 1	—	15	±3	38		45,7	63	2		+20,0	He.
42			6	—	16	4	7	41	47,6	48	53	2	+20,0	L.
43			7	—	18	4	30	13	47,9	47	17	5	+19,9	L.
44			6	—	18	4	37	43	48,0	60	57	41	+19,9	L.
45			6	—	20	4	53	43	48,0	61	15	5	+19,9	L.
46			6	—	20	5	0	14	48,0	49	7	17	+19,9	L.
47	13		6	—	20	5	1	24	49,7	65	25	10	+20,0	L.
48		VI, 35	C	—	20	5	6	24	49,3	60	23	58	+19,9	H
49			7	—	21	5	12	51	48,3	51	44	24	+20,0	L.
50	14	λ	5	—	21	5	12	53	48,[illegible]	53	25	5	+20,0	L.
51	15	κ	4	—	22	5	27	6	49,5	61	49	58	+20,0	Z. L.
52			7	—	23	5	38	11	49,5	59	26	50	+19,9	L.
53	16		6	—	23	5	44	26	50,3	65	39	5	+20,0	L.
54			7	—	23	5	49	43	48,4	46	47	9	+19,9	L.
55			7	—	24	5	52	59	48,4	49	52	3	+19,9	L.
56			7	—	24	6	3	19	49,6	60	54	51	+19,9	He.
57			6	—	25	6	11	50	50,3	63	30	40	+19,9	L.
58			7	—	25	6	16	58	48,9	53	4	14	+19,9	L.
59			6	—	25	6	20	1	49,9	59	13	43	+19,9	L.
60	17	ζ	4	—	26	6	29	21	48,9	52	48	1	+19,9	L.
61			6	—	27	6	45	47	48,5	46	19	6	+20,0	L.
62			7	—	27	6	47	31	48,5	50	8	30	+19,9	L.
63			6	—	28	6	57	35	48,7	45	51	45	+19,9	L.
64		II, 707	N	—	28	7	1	55	48,8	47	13	7	+19,9	H
65			6	—	28	7	3	25	48,7	48	29	40	+20,0	L.
66	18	α. *Schedir*, ð. 53″. V, 18	3	—	29	7	19	25	49,7	55	26	44	+19,9	Br.
67			7	—	30	7	31	19	49,0	51	6	23	+20,0	L.
68			6	—	30	7	35	0	51,7	65	3	9	+19,9	L.
69	19	ξ	6	—	31	7	45	25	49,1	49	25	7	+19,9	L.
70	20	π	6	—	32	8	7	37	48,9	45	55	55	+19,8	L.
71		VIII, 78	C	—	33	8	10	54	51,3	60	41	58	+19,8	H
72	21	r Cuſt. Meſs.	6	—	33	8	11	34	55,8	73	53	42	+19,8	L.
73	22	ο	6	—	34	8	25	25	49,1	47	11	32	+19,8	L.
74			6	—	34	8	30	27	50,0	54	7	48	+19,9	L.
75	23	21 Cuſt. Meſs.	6	—	35	8	41	7	56,4	73	45	22	+19,8	L.
76			7	—	35	8	44	44	41,1	54	22	11	+19,8	M
77			7	—	36	8	53	53	49,3	45	48	24	+19,8	L.
78			7	—	37	9	11	48	49,8	50	21	14	+19,8	L.
79	24	η. ð. 11″. III, 3	4	—	37	9	16	12	50,8	56	45	25	+19,8	Z. L.
80	25	ν	5	—	38	9	24	22	49,8	49	52	49	+19,7	L.
81			6	O.	39	9	37	57	49,7	46	40	42	+19,8	L.
82			6	—	40	9	54	33	50,0	50	25	16	+19,8	L.
83			6	—	40	10	4	15	50,4	50	29	6	+19,8	L.
84		I, 159	N	—	41	10	15	7	49,5	46	28	55	+19,7	H
85			6	—	41	10	19	36	52,4	60	1	44	+19,7	L.
86			6	—	42	10	26	50	50,3	48	31	42	+19,7	L.
87			8	—	42	10	33	28	51,0	55	13	42	+19,7	M
88			8	—	42	10	35	56	50,4	51	36	55	+19,7	M
89	26	ι. ν	7	—	43	10	49	18	52,1	57	53	29	+19,7	L.
90			5	—	44	10	53	39	50,3	47	34	14	+19,7	L.
91			6	—	44	10	57	32	50,1	50	14	1	+19,7	L.
92	27	γ	3	—	45	11	12	8	52,4	59	38	12	+19,7	Br.
93	28	2. ν	6	—	45	11	13	31	52,2	58	6	7	+19,7	L.
94			7	—	45	11	14	28	52,7	59	16	54	+19,6	L.
95			7	—	45	11	21	23	51,0	51	8	35	+19,7	L.
96			7	—	46	11	28	14	50,9	51	6	36	+19,7	L.
97			6	—	46	11	29	19	54,8	65	16	15	+19,6	L.
98			5	—	49	12	10	21	52,4	57	17	14	+19,7	L.
99			7	—	51	12	51	16	53,4	59	59	37	+19,5	M
100			8	—	52	12	58	50	53,8	57	57	42	+19,5	M
101			7	—	52	13	0	59	53,2	60	29	59	+19,5	M
102	29		6	—	53	13	10	0	52,5	63	58	14	+19,5	L.
103			7	—	53	13	18	16	51,5	49	58	34	+19,5	L.
104			6	—	54	13	30	5	54,9	61	44	41	+19,5	L.
105	30	μ	5	—	55	13	46	1	52,5	53	56	36	+19,5	L.
106			7	—	55	13	51	4	51,4	52	25	49	+19,5	L.
107			7	—	56	13	55	8	51,4	48	29	11	+19,4	L.
108		VIII, 64	C	—	56	14	0	1	54,7	60	31	8	+19,4	H
109			6	—	56	14	1	44	53,0	57	13	16	+19,5	M
110	31	ð. 20″. IV, 16	6	—	57	14	20	38	57,9	67	42	41	+19,4	L.
111			8	—	58	14	28	27	52,5	53	39	51	+19,4	M
112			7	—	58	14	31	40	52,0	50	39	10	+19,4	L.
113	32		6	—	59	14	43	5	56,3	63	57	13	+19,4	L.
114	33	θ	4	—	59	14	45	57	53,0	54	5	10	+19,4	Br.
115			7	I.	1	15	11	46	52,4	50	30	56	+19,4	L.
116		VII, 45	C	—	3	15	51	34	54,8	57	43	39	+19,3	H
117		VII, 42	C	—	7	16	42	10	54,8	57	11	25	+19,2	H
118	34	φ. ð. 12″. III, 23	6	—	8	16	54	59	54,9	57	10	49	+19,2	L.
119	35		7	—	8	16	59	7	57,8	63	36	44	+19,0	L.
120			7	—	8	17	5	15	57,2	62	28	7	+19,2	He.
121	35	ð. 43″. V, 81	7	—	9	17	9	5	57,9	63	56	53	+19,2	F
122	36	ψ. ð. 33″. V, 83.	5	—	12	18	0	54	60,5	67	5	3	+19,1	L.
123	37	δ	3	—	13	18	13	15	57,3	59	11	54	+19,1	Br.
124		VII, 48	C	—	16	18	58	20	58,6	62	17	13	+18,9	H
125			7	—	16	19	6	1	57,1	59	0	3	+18,9	L.
126	38	A Cuſt. Meſs.	5	—	17	19	9	7	63,1	69	13	57	+19,0	L.
127			7	—	18	19	30	33	59,0	62	21	38	+18,8	L.
128			6	—	18	19	32	36	62,0	67	22	39	+18,9	L.
129			6	—	19	19	50	48	59,0	62	37	12	+18,8	L.
130	39	χ	6	—	21	20	15	34	57,1	58	12	28	+18,8	L.
131	40	h. Cuſt. Meſs.	6	—	23	20	43	33	68,1	72	1	16	+18,8	L.
132			7	—	25	21	12	48	59,5	61	22	50	+18,7	L.
133	42	g. Cuſt. Meſs.	6	—	28	21	56	0	65,9	69	37	15	+18,6	L.
134	43	ω	6	—	28	21	56	55	63,5	67	1	39	+18,6	L.
135	41		6	—	28	22	0	14	62,2	71	36	49	+18,6	F
136		VII, 49	C	—	29	22	16	44	61,1	63	0	56	+18,5	H
137	44		6	—	30	22	29	41	58,9	59	22	29	+18,5	L.
138			8	—	30	22	33	10	63,3	66	32	12	+18,5	M
139		VII, 46	C	—	30	22	34	34	60,0	60	55	39	+18,5	H
140		VIII, 65	C	—	31	22	42	49	59,4	59	40	39	+18,5	H
141		VI, 31	C	—	33	23	10	49	60,0	60	13	39	+18,4	H
142			6	—	33	23	22	27	61,5	62	51	49	+18,5	L.
143	45	ε	3	—	40	25	3	29	61,4	62	41	2	+18,2	Br.
144	46	d	6	—	41	25	11	16	66,6	67	41	50	+18,2	L.
145			8	—	44	26	0	56	65,9	66	34	43	+18,0	M
146			7	—	45	26	11	11	65,9	66	1	44	+18,0	M
147	47	ð. 51″. V, 84. Rang.	6	—	46	26	25	24	82,3	76	18	50	+18,0	L.
148	48	e. Cuſt. Meſs.	5	—	46	26	28	2	70,2	69	55	58	+18,0	L.
149	50	f. Cuſt. Meſs.	4	—	47	26	40	48	70,6	71	26	54	+18,0	L.
150	49	18 Rangifer	6	—	47	26	44	14	79,2	75	8	46	+17,9	L.
151	52		7	—	48	27	3	14	64,5	63	55	49	+17,9	L.
152	51	43 Cuſt. Meſs.	6	—	48	27	3	40	76,9	73	36	44	+17,9	L.
153	53		7	—	48	27	6	23	64,1	63	25	6	+17,9	L.
154	54	44 Cuſt. Meſs.	6	—	52	28	3	21	72,5	70	36	20	+17,7	L.
155	55	ð. III, 4. I, 34	6	—	59	29	45	48	67,7	65	34	56	+17,4	L.
156		III, 695	N	II.	11	32	41	36	65,7	61	0	52	+16,9	H
157		*	4	—	13	33	13	6	71,0	66	29	50	+16,8	L.
158			7	—	17	34	22	41	72,3	66	31	47	+16,5	L.
159			7	—	19	34	51	11	72,7	66	37	48	+16,4	L.
160		VIII, 66	C	—	27	36	50	58	60,6	60	41	56	+16,8	H

Nº. 3. Fehlt nach Meſſier, allein es ſteht einer 1½° davon.
— - 10′ davon, ð. 21″. IV, 65.
— 14. ½° N. W. ð. 30″. V, 85.
— 29. ½° S. W. ð. I, 39.
— 56. Fehlt nach de la Lande *Pigott* hat ihn 1782. 1783. und 1784. vermiſst. (Transact. LXXVI. p. 203.)
— 81. ½° N. W. ð. I, 40 und ½° N. W. ð. 43″. V, 82.
— 114. 1½° W. ð. 24″. III, 66.
— 120. Fehlt nach de la Lande.
— 121. Fehlt nach de la Lande, und Herſchel hat ihn als Doppelſtern.
— 135. Fehlt nach de la Lande und Koch.
— 151. 153. Hiebey ſind nach de la Lande die Abweichungen beym Flamſteed verwechſelt.
— 155. Iſt dreyfach.

For 3 See 288 Cephei
119 ... 121 does not exist in that place, but is 119 ... 135 ... 142 in the 10th Col. add F.

Nº. 3. Manque ſelon Meſſier, mais il ſe trouve une à une diſt. de 1½°.
— - A une diſt. de 10′, ð. 21″. IV, 65.
— 14. ½° N. O. ð. 30″. V, 85.
— 29. ½° S. O. ð. I, 39.
— 56. Manque au ciel ſelon de la Lande. *Pigott* ne le trouve pas, en 1782. 1783. & 1784. (V. Tranſact. LXXVI, p. 203.)
— 81. ½° N. O. ð. I, 40 & ½° N. O. ð. 43″. V, 82.
— 114. 1½° O. ð. 24″. III, 66.
— 120. Manque ſuiv. de la Lande.
— 121. Manque ſelon d. l. Lande, mais Herſchel le met entre les étoiles doubles.
— 135. Manque ſuiv. de la Lande & Koch.
— 151. 153. Les declinaiſons de ces deux étoiles ſont confondues chez Fl. ſuiv. la recherche de de la Lande.
— 155. Eſt triple.

151-153 have by H.s ... page 208 ... the reverse with the British Catalogue.
152 Does not exist in this place ... See No. 57 Custos Mess.
157 in the 10th Col. add ...

B 2

Georg Heinrich Busse **Portrait of Caroline Herschel** **Hannover, Germany, 1847 P-144**

Caroline Herschel (1750–1848) was a German-English astronomer, the younger sister of astronomer Sir William Herschel and aunt of Sir John Herschel, with whom she worked throughout their careers. Her disfigured face, scarred by smallpox, and diminutive height, perhaps 4′ 3″ due to typhus, contributed to difficulties at home and to her spinsterhood. In 1772, while visiting his former home in Hanover, Germany, William rescued Caroline from a life of drudgery by returning to Bath, England, with sister in tow. Under William's tutoring, Caroline became a talented singer, but when William turned to astronomy, she had to give up her music and became instead his assistant, recording the coordinates of stars, writing out the details of a night's worth of observations, and copying scholarly papers.

Caroline made several significant contributions to astronomy, including the discovery of eight comets, one of which bears her name. No woman discovered more comets until Carolyn Shoemaker in the 20th century. For her index to Flamsteed's catalogue, as well as corrections to it, the Royal Astronomical Society awarded her its Gold Medal in 1828. The next woman to win this award, Vera Rubin, received her prize in 1996.

This portrait, done the year before her death, shows a very elderly Caroline pointing to a comet's path crossing the orbits of the planets. Although her diary reveals her displeasure at giving up her musical career to endure cold nights at the telescope, it also shows great pleasure at her discoveries of comets and nebulae. In her seventies, she even assisted John in his observations. Recent scholarship on her life provides fascinating details of the world's first female astronomer to draw a salary.

Saturn
Jupiter
Pallas
Vesta
Caroline Herschel
nach dem Leben gez. u. gest. v. G. Busse
geb. d. 16 ten März 1750.
Hannover 1847

Benjamin Tanner

"Astronomical Instruments. Dr. Herschel's Forty Feet Reflecting Telescope"

Philadelphia, United States, 1798 P-195

William Herschel built many variations of his 6.2-inch-diameter Newtonian reflector that he had used when discovering Uranus in 1781. For peering deeper and farther into the Universe, however, he relied on a 20-foot telescope that enabled him to see what no other contemporary astronomer could see. Realizing that an even larger mirror would reveal more of the Universe's secrets, he built a 40-foot telescope with a 48-inch-diameter mirror between 1785 and 1789. The world's largest telescope until its demise in 1840, it proved to be very difficult to use; its mirrors tarnished quickly, and the mechanism for moving the enormous tube was unwieldy, requiring several operators.

On the first night of its use in 1789, Herschel discovered a new moon of Saturn (Enceladus), followed by another (Mimas) the following month. So large and significant was this monster telescope that it was featured on local and Ordinance Survey maps and was itself a tourist attraction, especially for those on their way to visit the king at Windsor Castle, just a few miles away. King George III had sponsored its construction, and in the end granted £4,000 for it. The behemoth went completely unused after 1815. A surviving part of the tube is in storage at the National Maritime Museum in Greenwich, England, and one of the two original mirrors is on display at the London Science Museum.

The entire structure rests on a set of wheels that enables the telescope to point in any direction. A set of ropes and pulleys moves the tube up and down. Light enters the tube, is gathered by a mirror (housed near the shelter to the left), and comes to a focus near the open end of the tube, where the observer stands on the platform. Using a speaking tube, the observer shouts down to the scribe, armed with a notebook and light to take down notes in the shelter.

ASTRONOMICAL INSTRUMENTS.
PLATE XXXI.
Dr. HERSCHEL'S,
Forty Feet Reflecting
TELESCOPE.
G
O
F
N
E
H
I
L
K
M
R
C
B
A

William Herschel

Reflecting Telescope

Slough, England, c. 1788 G-29

William Herschel (1738–1822), a German musician, came to England in 1756. After settling in the resort town of Bath, his musical career slowly gave way to his passion for astronomy. Books by James Ferguson and Robert Smith taught him about astronomy and how to build his own telescopes. He soon built the most powerful telescopes in the world, seeing details that professional astronomers had been unable to observe.

On March 13, 1781, he used a telescope identical to this one to observe what he thought was a comet but turned out instead to be Uranus, the first planet discovered since antiquity. He soon gave up his musical career, and took a position as the King's astronomer. Herschel made a decent living from building and selling his telescopes; he made more than 400 of them over his lifetime. His financial situation improved greatly upon his marriage to a wealthy widow, with whom he had his only child, John, a noted astronomer in his own right and one of the most highly regarded scientists of the early-to-mid-19th century.

This instrument, a Newtonian reflector, consists of a primary mirror (6.2 inches in diameter) that bounces light to the top, where a mirror directs it to the eyepiece on the side. Herschel's ingenious mount – of which this is a reproduction – allows the eyepiece to stay at the same height regardless of the position of the telescope and the object in the sky.

John Flamsteed *Atlas Coelestis* **Paris, France, 1795 QB65 .F5 1795**

John Flamsteed (1646–1719), an English astronomer who served as the first Astronomer Royal, made the first sighting of Uranus in December 1690, although he thought it was a star. He refused to publish the results of his cataloguing of the heavens, a conflict that led Isaac Newton and Edmond Halley to publish that work, *Historia Coelestis Britannica*, without acknowledging Flamsteed's contributions. Flamsteed burned as many copies of it as he could, but his own version did not appear until 1725, after his death.

Flamsteed's original atlas included some of the constellations added by Gerard Mercator, Peter Plancius, Pieter Dirkszoon Keyser, and Frederick de Houtman in the 16th century, and by Johannes Hevelius in the 17th century. Several more constellations, created by Nicolas de Lacaille in the 18th century, were among the corrections included in the Paris edition of Flamsteed's atlas, pictured here.

After Herschel's discovery of Uranus in 1781, Jesuit astronomer Maximilian Hell proposed two new constellations, "le petit telescope de Herschel" (pictured here) and "le grand telescope de Herschel." These small and large examples of Herschel's telescopes were placed on either side of the point in the sky (located in the constellation Gemini) where Herschel had first observed Uranus. These constellations did not survive and are not connected to the now-standard constellation Telescopium, proposed by de Lacaille in 1752.

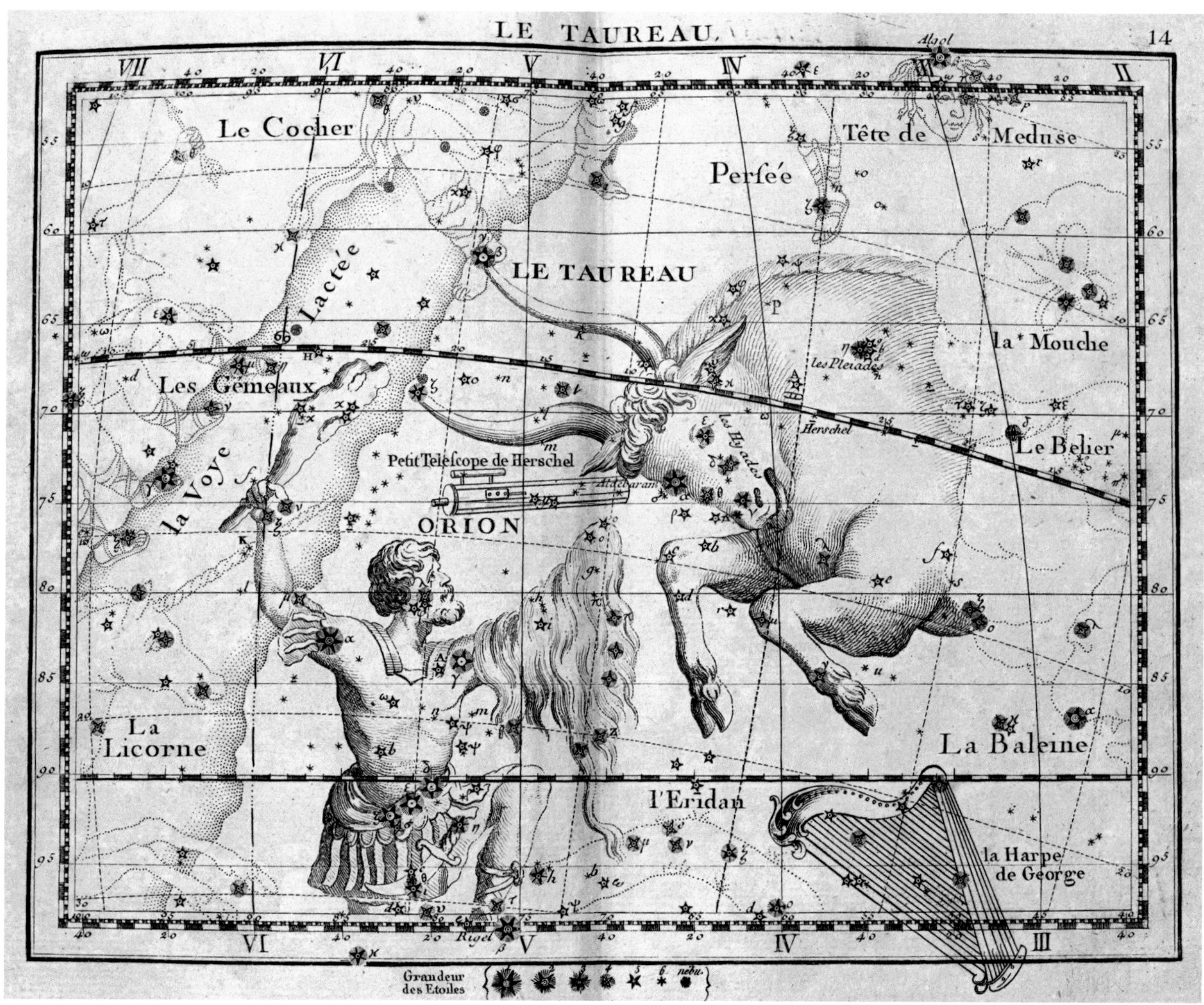
LE TAUREAU
14
Le Cocher
Tête de Meduse
Perſée
LE TAUREAU
la Voye Lactée
Les Gémeaux
la Mouche
les Pleiades
Herschel
Le Belier
Petit Teleſcope de Herſchel
Aldebaran
ORION
La Licorne
La Baleine
l'Eridan
la Harpe de George
Rigel
Algol
Grandeur des Etoiles

Mann & Ayscough

Refracting Telescope

London, England, 1743–1747 W-283

James Ayscough (d. 1759) designed and made scientific instruments, first as an apprentice to optician James Mann (1706–1756) and later as a partner with him from 1743 to 1747. Ayscough became known for his spectacles, telescopes, and microscopes, which he sold in his own London shop. His trade card indicates that, along with these and other optical items, such as prisms, camera obscuras, and magic lanterns, Ayscough sold thermometers, maps and globes, and hearing and speaking trumpets. Ayscough invented two techniques for providing two different objectives on one telescope. The earlier invention, an awkward one with two parallel lenses and holders, gave way to a much sturdier cube holding lenses in two perpendicular mounts, either of which could be rotated into place.

The main tube of this telescope has been covered in the tough skin of a stingray. The bumps have been shaved, leaving large white areas. The coloring of the spaces between these cells likely comes from the same rouge used to polish its lenses. The single-lens objective is housed in a brass cell, with a slider for protection. The three draws are covered in green vellum and have wooden aperture stops at the objective end, with milled brass ferrules at each ocular end. Inked, dotted extension marks indicate the proper distance to pull out each draw. As usual, the ocular unit has a three-lens eyepiece system. The numerous examples of telescopes with many of these features are often nearly indistinguishable from each other and difficult to date; in such cases, a signature might provide an important clue thanks to the efforts of scholars who have catalogued the often-changing business names and addresses.

John Kelly

Portrait of John Dollond from *The Life of John Dollond*

London, England, 1808

The problems of refracting telescopes, that is, telescopes using lenses to gather and focus light, were well known and well investigated since the time of Sir Isaac Newton. Despite Newton's declaration that these problems could not be solved, many astronomers and opticians nevertheless put their minds and hands to the task.

John Dollond (1706–1761), originally a silk weaver, taught himself Latin and Greek, as well as math and physics, especially optics. Having learned of the work of Chester Moor Hall and George Bass (as early as 1733) to produce achromatic lenses – that is, lenses that do not introduce colored halos into their magnified images – Dollond carried out further investigations in the 1750s. In 1752, he joined his first son, Peter (1730–1820), in the business of producing optical instruments, telescopes in particular.

Around 1757, John produced achromatic lenses consisting of a combination of flint glass and crown glass. After taking out a patent for this extremely important innovation, he did not take efforts to enforce it, but after his death, Peter took vigorous legal action against patent violators, putting many of them out of business.

OPTICKS.

Dollond

Refracting Telescope

London, England, c. 1760 W-181

When John Dollond (1706–1761) first successfully made an achromatic objective lens, he adjusted the structure of his eyepiece to include 5 rather than 3 lenses. Just a few years later, he made a further change, now using only 4 lenses. As a result, Dollond telescopes using the 5-lens eyepiece system are not only rare but datable to a brief period of time at the end of the 1750s. This example is unusual not only for its 5-lens eyepiece but for its presence in a wooden main tube.

Peter Dollond

Refracting Pocket Telescope

London, England, c. 1775 M-432

Pocket telescopes of even modest size or ornamentation are likely to launch many an interesting conversation, but one decorated in platinum leaf has little chance of failing. Its main tube, encased in tortoise shell, is inlaid with silver. The tubes and the objective-end aperture stop are made of pasteboard, with reflecting surfaces matte black. The single draw is covered in red leather and tooled in platinum, with an overall design based on leafy vines with a variety of single stamps of flowers and crescents. The lenses have silver or silvered-metal mounts attached to the pasteboard. The triple-lens objective is technological overkill but an illustration of the importance of the Dollond brand name as indicating a top-of-the-line instrument. Peter Dollond (1730–1820) had developed an achromatic triplet in 1765, and his rigid legal enforcement of any infringements ensured that this feature remained a desirable one. The single draw tube has a screw-on eyepiece, which has a small sliding lever that rotates the eyepiece lens from lens “1” (magnified and clear) to lens “2” (greater magnification but less clear).

Dolland

Refracting Telescope

London, England, 1800–1825 W-271

The Dollond brand became identified as the telescope for the serious astronomer, as well as for the amateur or casual observer, or gentlemen in search of a conversation piece. Peter Dollond's prosecution of anyone producing an achromatic lens in violation of his patent scared off many competitors, but one tactic proved an effective option: misspelling the name. Recent scholarly debate has not yet fully resolved how many variations of the "Dollond" name were legitimate spelling options and how many were intentionally fraudulent. Two of the more common deviations are "Dolon" and "Dolland", as seen here. The workmanship present in this example is high.

Dolland
London
Day or Night

Dollond

Refracting Telescope

London, England, 1843 A-62

By the middle of the 19th century, the Dollond patent on achromatic lenses had long expired. Nonetheless, ownership of a Dollond telescope still carried social weight, as the brand-name recognition continued throughout the 19th century. As the mass market for telescopes and other scientific instruments yielded stiff competition, fewer instruments received the elaborate ornamentation typical of earlier periods.

This example provides a wonderful exception. The hallmarks punched on each cap and tube confirm that it is made out of sterling silver that has been gold-plated. The main tube, push-on objective cap, and eyepiece are gold-embossed with raised scrollwork and foliate decoration. It also features two identical central decorations of telescope, comet, crescent moon, eight-pointed star, torch, drum, banners, and a chain. The lens cap, objective glass mount, and barrel are all punch-marked with five symbols indicating its maker, sterling quality, production in London, year, and duty. It has three draws, with the maker's inscription on the third. The rectangular domed case is covered in red Moroccan leather, with elaborate gold tooling.

Other examples of elaborately decorated telescopes by Dollond can be found in collections around the world, including the Imperial Palace of the Forbidden City in Beijing. The example illustrated here may have been produced for a Turkish noble.

Jesse Ramsden

Refracting Telescope

London, England, c. 1775 W-148

Jesse Ramsden (1735–1800), an English scientific-instrument maker, began an independent workshop around 1762. He made several improvements to telescopes, none more important than his dividing engine, which enabled him to divide a circle very accurately into its degree components and therefore allowed the user of instruments made by this machine to measure angles more accurately than ever before. His most successful precision instrument was a 5-foot diameter circle used in the Observatory at Palermo, Sicily. With it, Giuseppe Piazzi made the then-most-accurate catalogue of stars and also discovered the first asteroid, Ceres, on January 1, 1801. Ramsden earned a reputation for providing the highest-quality work at reasonable prices, often accompanied by a very long and frustrating wait despite the large workshop staff he employed.

In 1776, Ramsden married Sarah Dollond, the youngest sister of Peter Dollond. Ramsden seemingly took advantage of the opportunity to violate Dollond's patent, a situation that led to some animosity even before the couple's divorce. Many of Ramsden's telescopes are indistinguishable from those of Dollond, though the optics in this example are not of high quality.

Jesse Ramsden

Refracting Pocket Telescope

London, England, c. 1775   DPW-39

Although Jesse Ramsden (1735–1800) is best known for some of his large-scale telescopes and divided circles, this example shows his careful attention to detail and his flair for innovation. This brass refracting pocket telescope has many features seen in other instruments: screw-on brass end cap, achromatic objective lens, brass draw tube with eyepiece, with signature on the end ("Ramsden London"). This eyepiece has a rotating disc of four lenses, decreasing in size, with "1" being the largest. As a lens is rotated into place, its number appears in a separate circle cut in the eyepiece end plate. On the side of the eyepiece disk is a slot marked "O" and "S". By moving a tab extending from the slot from "O" to "S", a red shade is moved across the eye hole, allowing safe viewing of the Sun.

The telescope has an altazimuth mount, with a foot that slides into a slot on the base of the main tube. The pillar stand screws into a circular base atop a folding tripod. The nice innovation here is the stand that unscrews and folds for storage inside the main tube. The entire assembly is protected by a wooden case covered in sharkskin.

IV

Modern Telescopes

The telescopes of today and tomorrow will gather more information about this exciting mystery and will also undoubtedly suggest other riddles to solve.

The 19th century marked the golden age of refracting telescopes. In the 1810s and 1820s, lenses by Joseph von Fraunhofer graced Europe's leading observatories. By the 1860s, Alvan Clark's lenses were in high demand, especially in the American observatories that took a prominent role on the world's stage, culminating in the 40-inch lens at Yerkes Observatory in the 1890s. Problems due to the significant weight of ever-larger lenses and to the absorption of light passing through them directed attention again to reflecting telescopes, whose larger size enables them to gather more light than any lens. Since the construction of Yerkes, virtually all research telescopes use mirrors as their primary light-gathering tool.

A telescope increases the image size of the object being viewed, but even more importantly, it enlarges the size of the aperture or opening that gathers light coming from that object. Without that additional light, increased magnification means little. By the end of the 19th century, photography replaced the human eye, unveiling objects too faint for even assisted vision. The spectroscope (1859) revealed not only the component colors of incoming light but also lines that identified the chemical composition of the light source. Spectroscopy enabled astronomers to perceive color in a new and exciting way, providing clues about the composition and shape of the Milky Way and of the Universe itself.

The capabilities of 20th- and 21st-century telescopes have been expanded by increasing the sizes of their mirrors, by combining telescopes into large arrays, by extending their sensitivities beyond the optical spectrum and even into the realm of subatomic particles, and by incorporating sophisticated electronic devices and software. Computer-aided systems can change the shape of telescope mirrors to provide sharper images with better resolution. Modern telescopes are the central element of a complex technological system of specialized tools requiring extensive communications networks, data processing, and systems quality control.

Many of today's telescopes are located in nearly inaccessible locations. Distance from cities provides escape from light and other electromagnetic pollution, and high elevations place telescopes above much of the moisture and turbulence of the Earth's atmosphere. Space-based telescopes avoid all of Earth's atmospheric effects, some even traveling to distant worlds to view them from closer quarters.

Telescopic observations indicate that the Milky Way consists of about 100 billion stars, and that it is just one of perhaps 100 billion galaxies in the observable Universe. Telescopes provide ever-more convincing evidence for the Big Bang, the standard cosmological model today. One of the most exciting discoveries of the last few years suggests that the ordinary material and energy we encounter make up only 4% of the Universe; the rest consists of dark matter and dark energy, about which we know very little. This extraordinary finding, that most of the Universe is made up of completely unfamiliar and unknown material and energy, is at least as revolutionary as Copernicus's counterintuitive suggestion that the Earth orbits the sun. The telescopes of today and tomorrow will gather more information about this exciting mystery and will also undoubtedly suggest other riddles to solve. The curiosity that marked observers of centuries past remains as robust as ever, while the cosmos continues to inform, puzzle, and inspire us as we explore its wonders.

89

Antonio Zatta

"Planisferio Celeste Meridionale" from *Atlante Novissimo*

Venice, Italy, 1777 (published in 1779) P-42b

Astronomical observatories long predate the telescope. After its invention, though, the telescope inspired observatories on scales not seen before, both in size and in number. University observatories, such as those in Leiden (1633) and Utrecht (1642), the Netherlands, complemented the more famous national institutions in Copenhagen, Denmark (1637), Paris, France (1667), and Greenwich, England (1675). Others followed throughout Europe, including in Bologna (1725), Milan (1764), and Padua (1767–1777) in Italy.

Antonio Zatta (1757–1797), a leading northern Italian cartographer, engraved and published one of the most beautiful late-18th-century atlases, the *Atlante Novissimo*. In a set of planispheres illustrating the stars and constellations of both northern and southern hemispheres, Zatta included depictions of what he considered the most significant observatories of the day: Greenwich, Paris, Copenhagen, Kassel (Germany), Padua, Pisa, Bologna, and Milan. Zatta's selection certainly demonstrates some regional bias.

Astronomy and its observatories nicely illustrated the Age of Reason, also known as the Enlightenment, which spread through Europe during the 18th century. Common sense suggested that the Sun orbits the Earth, but reason showed that the Earth orbits the Sun. Observatories served as the temples of reason, with their architecture often making that point in elegant ways.

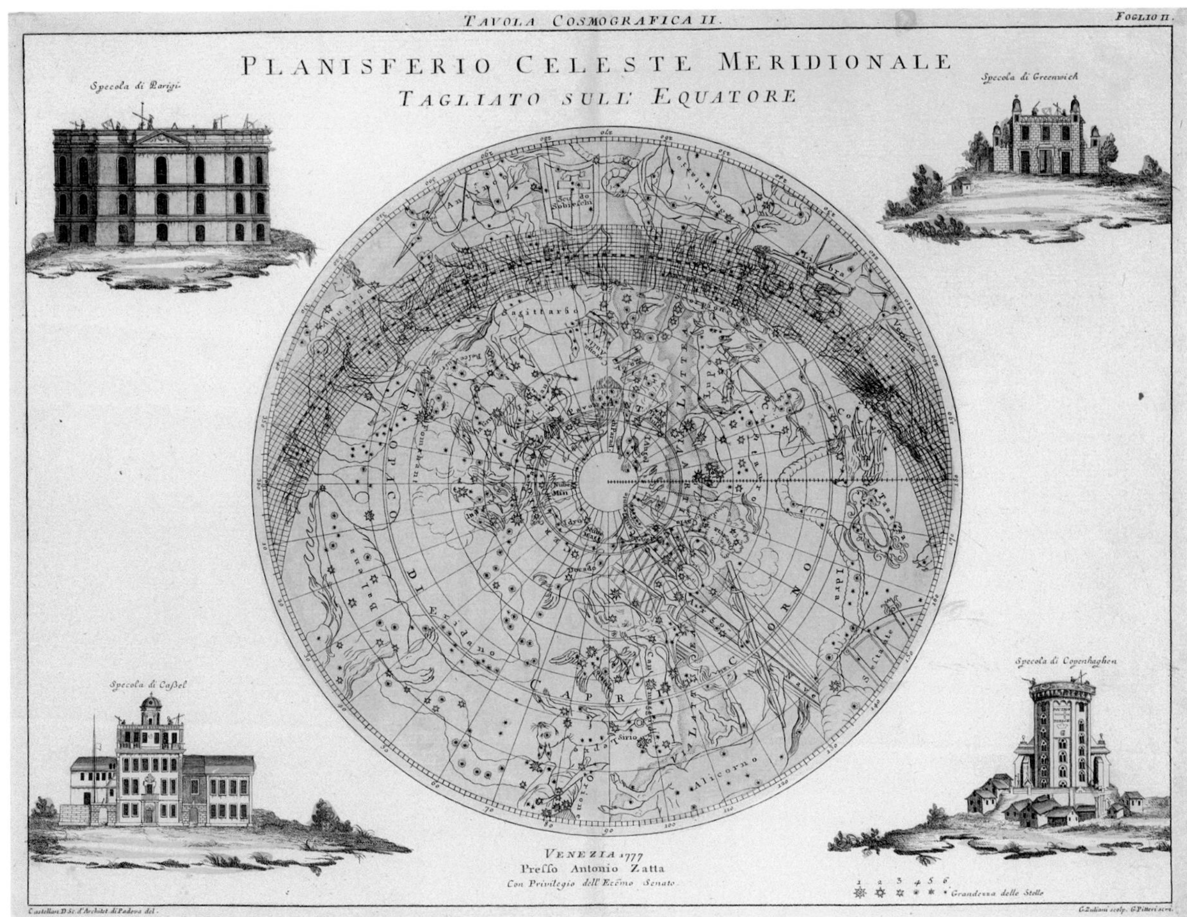
TAVOLA COSMOGRAFICA II.
FOGLIO II.
PLANISFERIO CELESTE MERIDIONALE
TAGLIATO SULL' EQUATORE
Specola di Parigi
Specola di Greenwich
Specola di Caßel
Specola di Copenhaghen
VENEZIA 1777
Presso Antonio Zatta
Con Privilegio dell' Eccmo Senato
1 2 3 4 5 6 Grandezza delle Stelle

90

Joseph von Fraunhofer

Bestimmung des Brechung

Munich, Germany, 1817

Joseph von Fraunhofer (1787–1826), a German optician and astronomer, is remembered for his careful illustration of the dark solar absorption spectrum lines (the so-called Fraunhofer lines). Orphaned at a young age, he was rescued in two ways by the future King of Bavaria, Maximilian I: first, literally, from the wreckage of a workshop building that collapsed with Fraunhofer in it, and second, from a life of drudgery when the prince recognized Fraunhofer's talents and provided him with an excellent education. While testing his optical glass, Fraunhofer encountered the absorption lines that he put to use for measuring its optical properties, but he never investigated the causes of those absorption lines.

Fraunhofer studied and measured the wavelengths of more than 570 solar absorption lines. Although modern observations of sunlight can detect many thousands of lines, Fraunhofer's discovery went far beyond earlier casual and dismissive observations, such as those by Newton. Many visitors came to visit Fraunhofer, including John Herschel, son of William Herschel and an astronomer in his own right. John learned of spectral lines from Fraunhofer and suggested their potential significance in the 1820s, but never pursued it. In 1859, Robert Bunsen and Gustav Kirchhoff discovered that patterns of spectral lines are associated with each chemical element or compound and deduced that the dark lines Fraunhofer observed in the solar spectrum were caused by absorption by those elements in the upper layers of the Sun.

This image is the first printed illustration of the lines that would completely revolutionize astronomy in the 19th century, allowing astronomers to know the composition and motion of stars and gases anywhere in the visible Universe.

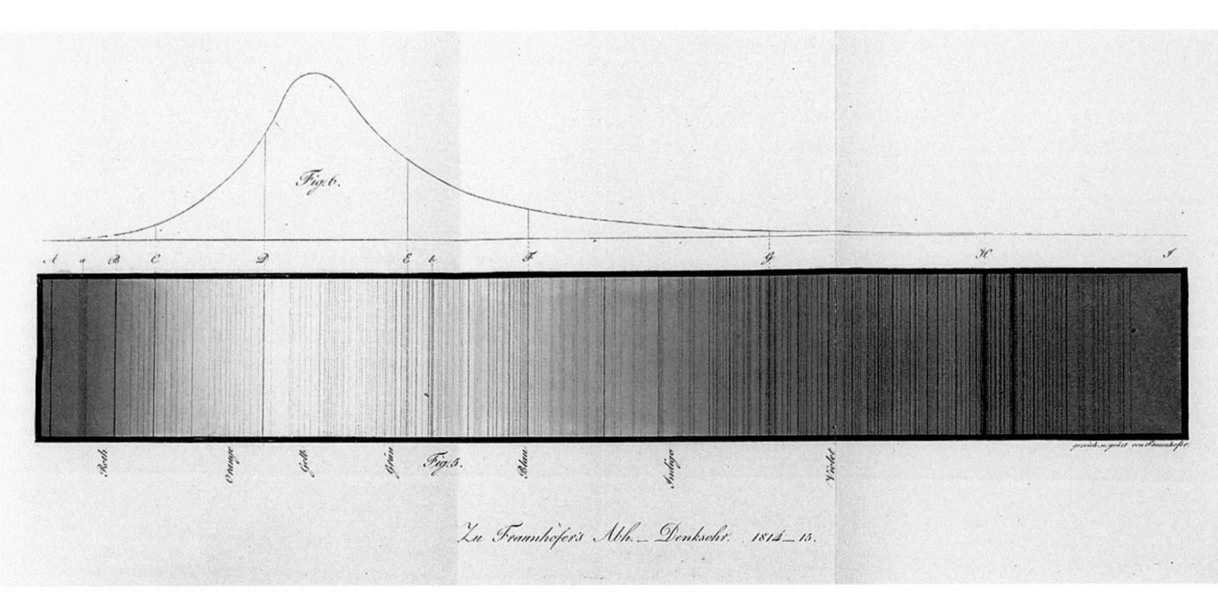
Fig. 6.
A
a
B
C
D
E
b
F
G
H
I
Roth
Orange
Gelb
Grün
Fig. 5.
Blau
Indigo
Violet
Zu Fraunhofer's Abh. – Denkschr. 1814–15.

91

Joseph von Fraunhofer

Refracting Telescope

Benedictbeuern, Germany, 1809–1814 A-407

Joseph von Fraunhofer (1787–1826) survived several tragedies in his early life to become the most important maker of telescopes in the early 19th century. Ironically, exposure to the hazardous chemicals and vapors associated with glassmaking likely contributed to Fraunhofer's early death. His education, provided through the generosity of the future king of Bavaria, led Fraunhofer to the Optical Institute at Benedictbeuern, a secularized Benedictine monastery just outside of Munich that was a leading center for making optical glass. He was known above all for making the world's best optical glass and achromatic telescope objectives, which found homes in the leading observatories of the day and led to many important discoveries.

At the age of only 31, Fraunhofer became the director of the Optical Institute and developed the recipes and processes for making the highest-quality optical glass. One of the most important secrets was learned from Pierre Guinand, that the cooling glass must be stirred to remove bubbles. Astronomers from around Europe came to visit Fraunhofer to learn his secrets, though they went away empty-handed. Fraunhofer's name became synonymous with the best telescopes that could be bought, but their superiority was mostly limited to the upper-end market. Many instruments bearing his name never benefited from the master's hand and are little better than other brands.

This example is signed on the eye tube: "Utzschneider Reichenbach und / Fraunhofer in Benedictbeurn". The main tube is brass with a plain brown leather grip and two ferrules. It has one main draw tube and an eyepiece tube; the brass parts screw together or fit by friction. The telescope has an original objective lens and a four-lens eyepiece.

Fraunhofer in Benedictbeurn

92

Friedrich Georg Wilhelm Struve

Description de l'Observatoire

St. Petersburg, Russia, 1845

The Struve family of astronomers extended over five generations from the 18th through the 20th centuries. Friedrich Struve (1793–1864), in the second generation, was the first of several family members associated with the Dorpat Observatory (equipped with a Fraunhofer telescope) in Tartu, Estonia, and with St. Peterburg's Pulkovo Observatory, which he founded and directed. His most important work involved the measurement and cataloguing of double stars.

From the middle to the end of the 19th century, astronomical telescopes grew in size and became housed in increasingly sophisticated and technically designed observatories. In this volume, Struve provides a catalogue of the important examples located throughout Europe. Note the sturdy foundational structures that provide needed stability for the main telescope located just under the upper dome. Just as noteworthy, though, are the architectural elements that strongly resemble the features of classical temples and cathedrals. The 19th-century observatory stood as a monument not only to astronomy but more broadly to science, engineering, and the triumphs of reason.

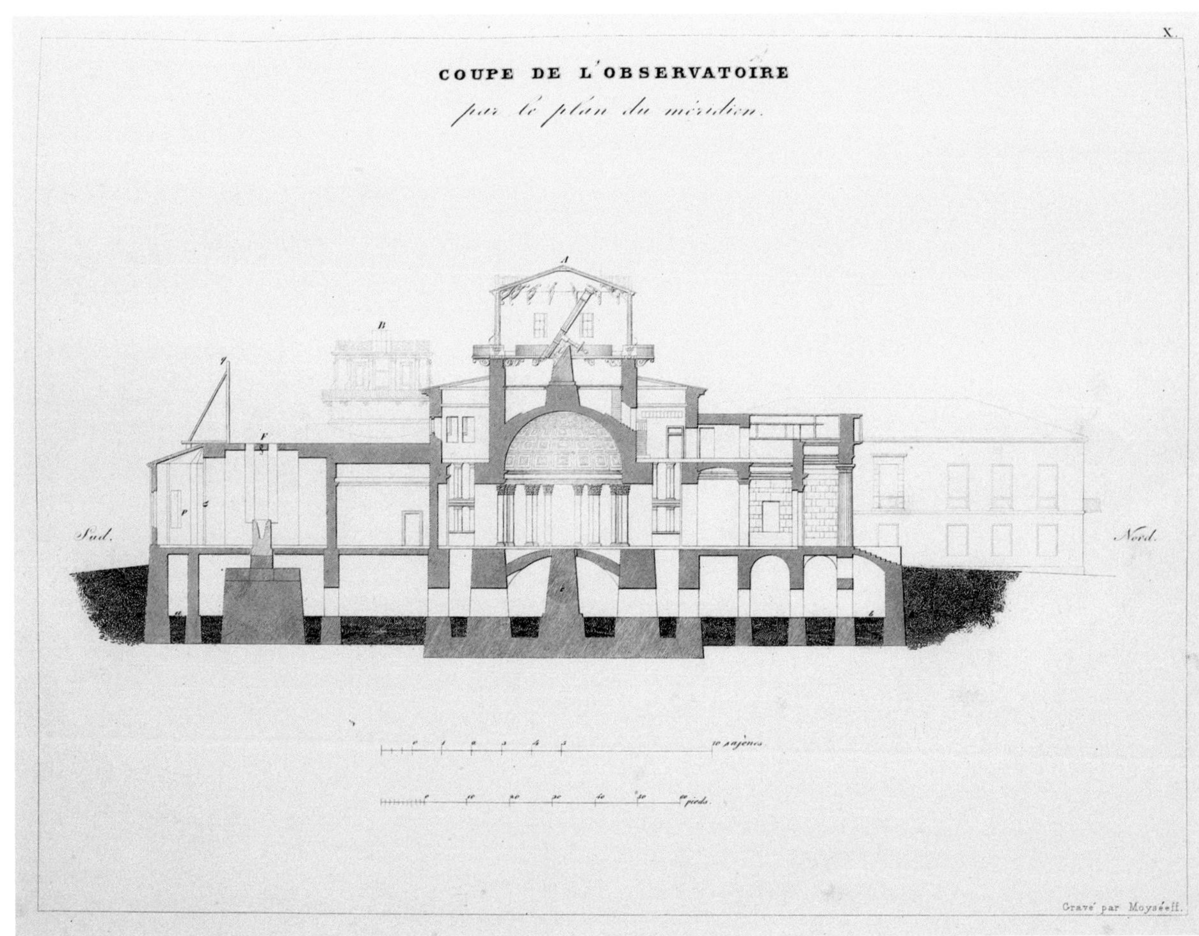
X.
COUPE DE L'OBSERVATOIRE
par le plan du méridien.
Sud.
Nord.
10 sajènes
pieds.
Gravé par Moyséeff.

93

William Huggins

On the results of spectrum analysis applied to the heavenly bodies

London, England, 1866

Shortly after Robert Bunsen (1811–1899) and Gustav Kirchhoff (1824–1887) demonstrated the significance of spectroscopy in 1859, William Huggins (1824–1910) adapted his private observatory in London to conduct spectroscopic observations of diverse celestial objects. He was assisted by his neighbor, chemist William Miller (1817–1870), and later by his wife, Margaret (1848–1915), whom he married in 1875. Huggins was able for the first time to distinguish between celestial gas clouds (such as the Orion nebula, which has a bright-line emission spectrum) and stars or collections of stars (such as the Andromeda nebula or galaxy, which has a dark-line absorption spectrum). These findings provided a tantalizing suggestion that there might be galaxies outside the Milky Way, but with no way to determine distances to more than just a few stars, it was still impossible to know if the model of the Universe should include more than just one galaxy.

This volume consists of the printed copy of a lecture by Huggins given at the annual meeting of British Association for the Advancement of Science in Nottingham (1866). Each illustration, an actual photograph pasted onto a text page, is taken from one of the series of glass plates used to project the illustrations used by Huggins during his lecture. These photographs are of drawings, not direct photographs of spectra themselves. The image shown, one of the first published images of a stellar spectrum, shows that stars have readily distinguishable spectral fingerprint patterns. It also demonstrates that already within a few years of the invention of this new technique, it was possible to find and separate the spectral patterns of closely situated double stars.

Orange.
Blue.
W. Huggins. Fecit

94

Camille Flammarion ***Terres du Ciel*** **Paris, France, 1884**

Nicolas Camille Flammarion (1842–1925), a French astronomer, wrote prolifically about astronomy and other scientific topics, including spiritualism and reincarnation. He was an enthusiastic supporter of the idea of the plurality of worlds, the notion of the presence of life elsewhere in the Universe. Many of his books featured drawings of intelligent creatures on other planets. His first book defending this position went into more than 30 editions in several languages. Although he was rightly regarded by his contemporaries as eccentric, his writings were enormously popular and brought astronomy to public attention and popular imagination. Perhaps his most famous illustration shows a man poking his head through the starry firmament and observing what lies on the other side. (Although this image appears for the first time in his 1888 volume on popular meteorology, it is often used to illustrate the supposedly simplistic worldview of the Middle Ages.)

In his *Terres du Ciel* (*Worlds of the Heavens*), Flammarion continues to put forth his view that the Moon and Mars, as well as other planets, support intelligent life. Elsewhere, Flammarion explicitly compares the surfaces of Mars and Earth; here, his illustration of Martian topography reveals his belief in their similarity. This follows up on the 1877 observations of Mars by Italian astronomer Giovanni Schiaparelli (1835–1910), who described "canali" on Mars, using a word that can be translated as (natural) channels or (artificial) canals. The ambiguity led many to promote the notion of life existing on Mars. Flammarion believed that the "canals" of Mars were probably built by an advanced civilization, one of the topics of his *La planète Mars et ses conditions d'habitabilité* (*The planet Mars and the conditions of its habitability*, 1888) though he presented his views here in prose less florid than in his earlier work. He allowed that the *canali* might be due to geological processes, but, like William Herschel a century earlier, he concluded that "the actual habitation of Mars by a race superior to our own is in our opinion very probable."

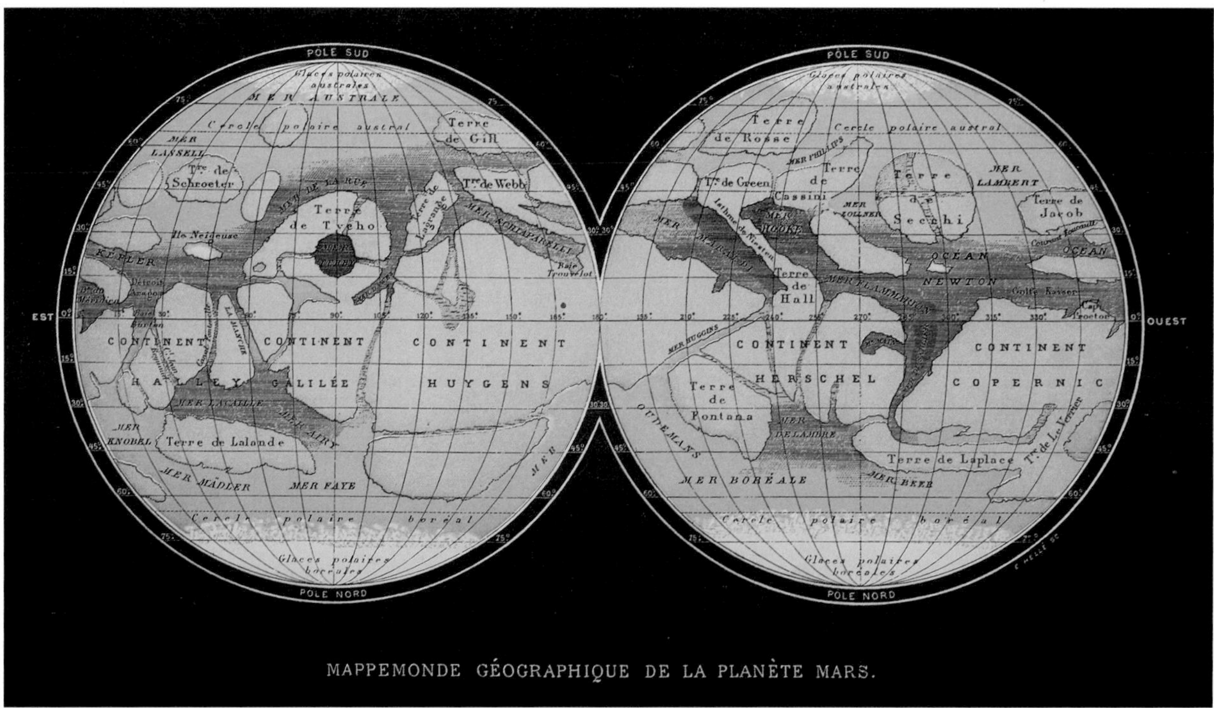

MAPPEMONDE GÉOGRAPHIQUE DE LA PLANÈTE MARS.

95

Alvan Clark & Sons **Refracting Telescope** **Cambridge, Massachusetts, 1864 G-33**

Alvan Clark (1804–1887), a portrait painter turned telescope maker, built many of the best (and successively, the five largest) refracting telescopes in the world, including the 40-inch at Yerkes Observatory, which is still the world's largest. The first of these world-recordholders employed an 18.6-inch achromat objective lens. While his son Alvan Graham Clark was testing it, he discovered the white dwarf companion of Sirius. That lens was housed in this tube and intended for the University of Mississippi, but the intrusion of the Civil War kept it north of the Mason-Dixon line and eventually brought it to the old University of Chicago and then to the observatory at Northwestern University in Evanston, Illinois. The lens is still used there, but the tube and mount came to the Adler when it opened in 1930.

The walnut veneer tube has two counterweight poles running along the top and bottom; these have iron balls at the eyepiece ends. Side-mounted on the tube, with turning handles at the eyepiece end (and therefore convenient to the observer), are a worm-screw mechanism connected to the horizontal axis for fixing the telescope in a certain position and another worm screw and gear train for slow motion and fine altitude adjustments. The tube has four brass ferrules, at the objective end and along the body of tube; a fifth for the eyepiece end is missing.

For much of its display history, the tube was tarred to protect it from the dry air of Chicago winters. In 1999, prior to installation in the Adler's newly expanded building, it underwent conservation treatments that revealed the beautiful craftsmanship of the telescope furniture.

96

Western B. N. Engraving Co.

"The Chicago Astronomical Society" Certificate of Membership

Chicago, Illinois, 1875 P-191

This blank certificate indicates membership to the Chicago Astronomical Society (CAS), the oldest such group in the United States and likely the second oldest in the world. It was founded in 1862 to improve the young city's reputation as a cultural and scientific center. It originally brought together both professional and amateur astronomers to promote astronomical knowledge. Perhaps its most significant accomplishment was the construction of the Dearborn Observatory and the purchase of the 18.6-inch Alvan Clark refractor, at that time the largest in the world. Still active today, CAS carries out its mission by maintaining an observatory, hosting diverse public events, including observing sessions, and sponsoring regular meetings and lectures.

THE Chicago Astronomical Society

THE GREAT EQUATORIAL TELESCOPE

FOUNDED NOVEMBER, 1862.
INCORPORATED FEBRUARY 19, 1867.

REORGANIZED AFTER THE GREAT FIRE JULY 9, 1875.

FOCAL LENGTH 23 FEET.

DEARBORN OBSERVATORY:
Tower built 1864–65. Telescope placed in position, March 1866.
Dome rebuilt in the Spring of 1875. Total cost of Buildings & Instruments $ 70,000.

It is Hereby Certified, that, ______________________

is a Life ______________________ of

THE CHICAGO ASTRONOMICAL SOCIETY

Attest

______________________ President.

______________________ Vice President.

______________________ Treasurer.

______________________ Secretary.

Made by ALVAN CLARK & SONS.
DIAMETER OF OBJECT GLASS, 18 ¾ INCHES.
COST $ 18,187.00 HEIGHT OF FLOOR 66 FEET.

Western B. N. & Engraving Company.
CHICAGO.

97

Chicago Tribune

"The Graphic Weekly"

Chicago, Illinois, May 15, 1932 P-49

By May 15, 1932, when this article appeared in "The Graphic Weekly" section of the Sunday edition of the *Chicago Tribune*, the Yerkes 40-inch refracting telescope was no longer the world's largest telescope. Then, as now, it still held the record as the largest refractor, but it had been overtaken in size by the 60-inch reflector on Mt. Wilson near Los Angeles in 1908, and again by the 100-inch reflector there in 1917. Even so, the Yerkes telescope, unveiled at the Columbian Exposition in Chicago in 1893, still featured prominently in public imagination.

George Ellery Hale had convinced Charles T. Yerkes, who made his fortune by financing Chicago's electric railway system, to fund the world's largest refracting telescope and thereby earn lasting recognition. The preeminent telescope maker Alvan Graham Clark, son of Alvan Clark who made the Dearborn telescope, created the lenses for this instrument, located in Williams Bay, Wisconsin, on the shores of Geneva Lake.

This article highlighted the Yerkes telescope as a marvel of technical engineering. Today, while remaining the supreme example of the telescope-making craft of the 19th century, it functions as an extraordinary special-purpose optical observing instrument. Along with its stunning architectural complex, the Yerkes instrument provides a lasting testimony to the great era of refracting telescopes.

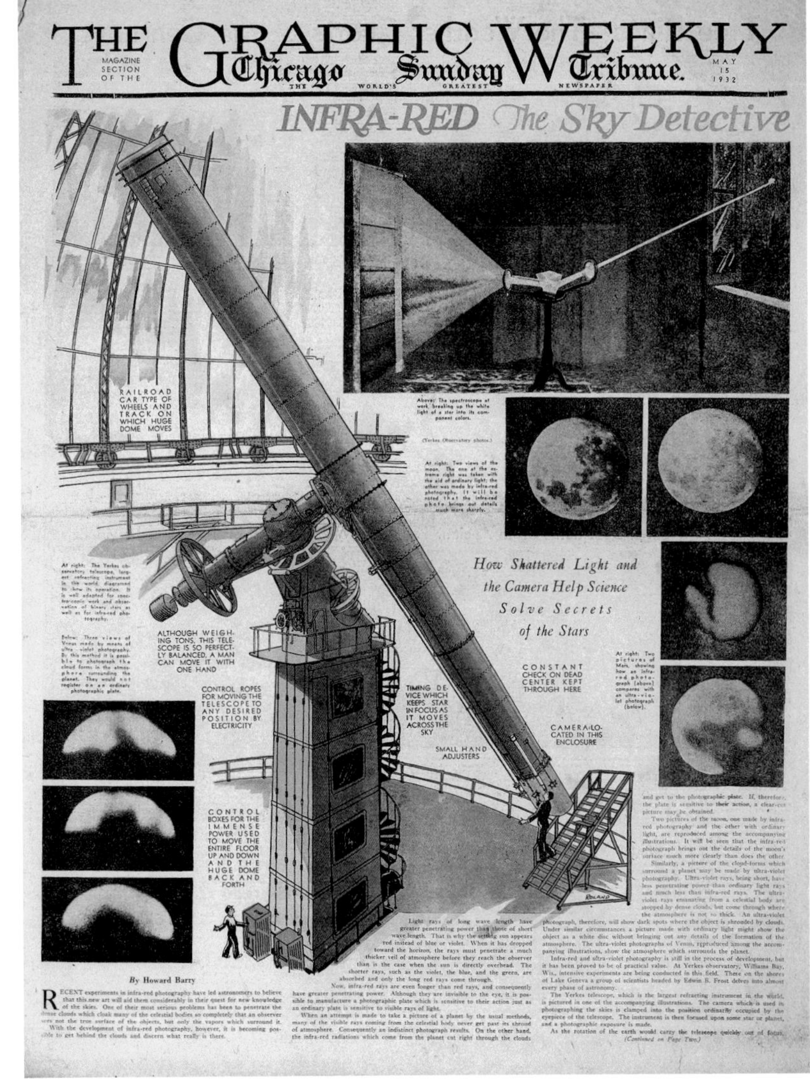

THE GRAPHIC WEEKLY

MAGAZINE SECTION OF THE Chicago Sunday Tribune. THE WORLD'S GREATEST NEWSPAPER

MAY 15 1932

INFRA-RED The Sky Detective

Above: The spectroscope at work, breaking up the white light of a star into its component colors.

(Yerkes Observatory photos.)

At right: Two views of the moon. The one at the extreme right was taken with the aid of ordinary light; the other was made by infra-red photography. It will be noted that the infra-red photo brings out details much more sharply.

At right: The Yerkes observatory telescope, largest refracting instrument in the world, diagramed to show its operation. It is well adapted for spectroscopic work and observation of binary stars as well as for infra-red photography.

Below: Three views of Venus made by means of ultra-violet photography. By this method it is possible to photograph the cloud forms in the atmosphere surrounding the planet. They would not register on an ordinary photographic plate.

RAILROAD CAR TYPE OF WHEELS AND TRACK ON WHICH HUGE DOME MOVES

ALTHOUGH WEIGHING TONS, THIS TELESCOPE IS SO PERFECTLY BALANCED, A MAN CAN MOVE IT WITH ONE HAND

CONTROL ROPES FOR MOVING THE TELESCOPE TO ANY DESIRED POSITION BY ELECTRICITY

CONTROL BOXES FOR THE IMMENSE POWER USED TO MOVE THE ENTIRE FLOOR UP AND DOWN AND THE HUGE DOME BACK AND FORTH

TIMING DEVICE WHICH KEEPS STAR IN FOCUS AS IT MOVES ACROSS THE SKY

SMALL HAND ADJUSTERS

CONSTANT CHECK ON DEAD CENTER KEPT THROUGH HERE

CAMERA LOCATED IN THIS ENCLOSURE

ROLAND

How Shattered Light and the Camera Help Science Solve Secrets of the Stars

At right: Two pictures of Mars, showing how an infra-red photograph (above) compares with an ultra-violet photograph (below).

By Howard Barry

RECENT experiments in infra-red photography have led astronomers to believe that this new art will aid them considerably in their quest for new knowledge of the skies. One of their most serious problems has been to penetrate the dense clouds which cloak many of the celestial bodies so completely that an observer sees not the true surface of the objects, but only the vapors which surround it.

With the development of infra-red photography, however, it is becoming possible to get behind the clouds and discern what really is there.

Light rays of long wave length have greater penetrating power than those of short wave length. That is why the setting sun appears red instead of blue or violet. When it has dropped toward the horizon, the rays must penetrate a much thicker veil of atmosphere before they reach the observer than is the case when the sun is directly overhead. The shorter rays, such as the violet, the blue, and the green, are absorbed and only the long red rays come through.

Now, infra-red rays are even longer than red rays, and consequently have greater penetrating power. Although they are invisible to the eye, it is possible to manufacture a photographic plate which is sensitive to their action just as an ordinary plate is sensitive to visible rays of light.

When an attempt is made to take a picture of a planet by the usual methods, many of the visible rays coming from the celestial body never get past its shroud of atmosphere. Consequently an indistinct photograph results. On the other hand, the infra-red radiations which come from the planet cut right through the clouds and get to the photographic plate. If, therefore, the plate is sensitive to their action, a clear-cut picture may be obtained.

Two pictures of the moon, one made by infra-red photography and the other with ordinary light, are reproduced among the accompanying illustrations. It will be seen that the infra-red photograph brings out the details of the moon's surface much more clearly than does the other.

Similarly, a picture of the cloud-forms which surround a planet may be made by ultra-violet photography. Ultra-violet rays, being short, have less penetrating power than ordinary light rays and much less than infra-red rays. The ultra-violet rays emanating from a celestial body are stopped by dense clouds, but come through where the atmosphere is not so thick. An ultra-violet photograph, therefore, will show dark spots where the object is shrouded by clouds. Under similar circumstances a picture made with ordinary light might show the object as a white disc without bringing out any details of the formation of the atmosphere. The ultra-violet photographs of Venus, reproduced among the accompanying illustrations, show the atmosphere which surrounds the planet.

Infra-red and ultra-violet photography is still in the process of development, but it has been proved to be of practical value. At Yerkes observatory, Williams Bay, Wis., intensive experiments are being conducted in this field. There on the shores of Lake Geneva a group of scientists headed by Edwin B. Frost delves into almost every phase of astronomy.

The Yerkes telescope, which is the largest refracting instrument in the world, is pictured in one of the accompanying illustrations. The camera which is used in photographing the skies is clamped into the position ordinarily occupied by the eyepiece of the telescope. The instrument is then focused upon some star or planet, and a photographic exposure is made.

As the rotation of the earth would carry the telescope quickly out of focus,

(Continued on Page Two.)

98

William Morgan, University of Chicago, Yerkes Observatory

Milky Way Model

Williams Bay, Wisconsin, 1951 NS-97-1

From the first telescopic observations in 1608, people wondered about the size and shape of the Milky Way. Many people over the ensuing years speculated that the Milky Way has spiral arms, but it took until 1951 for confirmation. In the previous decade, Walter Baade had noted an important feature of spiral galaxies: the older (redder) stars are typically located in the galaxy's central regions, whereas the younger (blue-white) stars are located (along with ionized hydrogen) in the spiral arms. In 1951, William Morgan collaborated with his colleagues at Yerkes Observatory to determine the distances to various nearby blue stars and hydrogen regions. When they charted this information, they had mapped portions of three of the Milky Way's spiral arms, which he named Perseus, Orion, and Sagittarius, after the constellations containing prominent stars of these arms. Morgan's announcement, illustrated with a model using cotton balls to show these spirals, drew an animated response and a standing ovation at that year's meeting of the American Astronomical Society. (The results were published in Morgan, Sharpless, & Osterbrock, "Some features of Galactic Structure in the Neighborhood of the Sun," in *The Astrophysical Journal*, Volume 57 [1952].) Later radio observations improved this model, leading to the thorough map of the barred spiral structure of the Milky Way galaxy that we use today.

60°
UNOBSERVED

Edmund Scientific Corporation

Satellite Tracking Telescope

Barrington, New Jersey, c. 1955 A-332

Since the invention of the telescope in 1608, it has been used as a tool for astronomy, military spying, and surveying, and also as a fun or educational toy. In the 1950s, a new purpose emerged. The International Geophysical Year (1957–1958) enlisted the collaboration of 67 nations in a program incorporating eleven sciences and provided a novel suggestion to track artificial satellites orbiting the Earth, using that data to map the Earth's surface. In the United States, in the midst of a Cold War with the Soviet Union, participants in Operation Moonwatch leapt into action when Sputnik launched in 1957.

Teams of observers formed a line enabling them to watch large swaths of the sky, making sure that a satellite did not escape their notice. The desire to contribute to scientific knowledge was complemented by active involvement in the struggle against a political and social enemy. Today, by contrast, international scientific collaborations have replaced Cold War era competition, particularly in space flight.

Many Moonwatch observers purchased a telescope, shown here, specially designed for this project. The uncertainty of the appearance of a satellite such as Sputnik meant that it was difficult to locate it. The optics of this telescope provides a very large field of view, enabling the observer not only to locate a satellite but also to time its progress across the sky; this information was vital for the purposes of Operation Moonwatch. To make it possible and painless for the viewer to observe for a long period of time, the telescope points down and uses a mirror. To cut down on eye strain and avoid contact with cold metal, a rubber ring around the eyepiece provides relief.

Edscorp
SATELLITE SCOPE
EDMUND SCIENTIFIC CO.
SERIAL NO.
117

Bibliography

Partially Annotated Bibliography of Major Works

Major Publications

Biagioli, Mario. *Galileo's Instruments of Credit.* Chicago: University of Chicago Press, 2006.

Bredekamp, Horst. *Galilei der Künstler.* Berlin: Akademie Verlag GmbH, 2007. Soon to appear in English translation.

Hoskin, Michael. *The Herschel Partnership, as viewed by Caroline.* Cambridge: Science History Publications, 2003. Other volumes by Hoskin on the Herschel family: *The Herschels of Hanover* (2007), *Caroline Herschel's Autobiographies* (2003), *William Herschel and the Construction of the Heavens* (1963), *William Herschel, Pioneer of Sidereal Astronomy* (1960).

King, Henry. *The History of the Telescope.* Cambridge, Mass.: Sky Pub. Corp., 1955. This is the best detailed history in English, soon to be complemented by the English translation of Riekher's volume.

McCray, Patrick. *Giant Telescopes: Astronomical Ambition and the Promise of Technology.* Cambridge, Mass.: Harvard University Press, 2006.

Riekher, Rolf. *Teleskope und ihre Meister, 2nd ed.* (*Telescopes and their Masters*). Berlin: Verlag Technik, 1990. An English translation should appear in 2009.

Strano, Giorgio, ed. *Galileo's Telescope: The Instrument that Changed the World.* Florence: Giunti, 2008.

Van Helden, Albert. "The Invention of the Telescope." *Transactions of the American Philosophical Society* 67:4 (1977): 1–67. Reprinted in 2008 under the same title and publisher.

Watson, Fred. *StarGazer: The Life and Times of the Telescope.* Cambridge: Da Capo Press, 2004.

Willach, Rolf. *The Long Route to the Invention of the Telescope.* Philadelphia: American Philosophical Society, 2008.

Other Related Studies

Andersen, Geoff. *The Telescope: Its History, Technology, and Future.* Princeton: Princeton University Press, 2007.

Bell, Louis. *The Telescope.* McGraw-Hill: York, Pennsylvania, 1922. This classic has appeared in several Dover editions.

Biagioli, Mario. *Galileo, Courtier.* Chicago: University of Chicago Press, 1993.

Bolt, Marvin and Michael Korey. "Trumpeting the Tube: A Survey of Early Trumpet-shaped Telescopes," 146–163 in Hamel and Keil, *Der Meister und die Fernrohre.*

Dunn, Richard. *The Telescope: A Short History.* London: National Maritime Museum Publishing, 2009.

Hamel, Jürgen, and Inge Keil, eds. *Der Meister und die Fernrohre: Das Wechselspiel zwischen Astronomie und Optik in der Geschichte.* Frankfurt: Verlag Harri Deutsch, 2007.

Hirshfeld, Alan. *Parallax: The Race to Measure the Cosmos.* New York: Holt Paperbacks, 2002.

Ilardi, Vincent. *Renaissance Vision from Spectacles to Telescopes.* Philadelphia: American Philosophical Society, 2007.

Jackson, Miles. *Spectrum of Belief: Joseph von Fraunhofer and the Craft of Precision Optics.* Cambridge, Mass: The MIT Press, 2000.

McConnell, Anita. *Jesse Ramsden: London's Leading Scientific Instrument Maker.* Burlington, Vermont: Ashgate, 2007.

Panek, Richard. *Seeing and Believing: A Short History of the Telescope and How We Look at the Universe.* New York: Diane Publishing Co., 1998.

Reeves, Eileen. *Galileo's Glassworks: The Telescope and the Mirror.* Cambridge, Mass.: Harvard University Press, 2008.

_____. *Painting the Heavens.* Princeton: Princeton University Press, 1997.

Rudd, M.E., Duane H. Jaecks, Rolf Willach, Richard Sorrenson, and Peter Abrahams, "New light on an old question: Who invented the achromatic telescope?" *Journal of the Antique Telescope Society* 19 (2000): 3–12.

Rudd, M.E. "Chromatic Aberration of eyepieces in early telescopes." *Annals of Science* 64: 1 (2007): 1–18.

Van Helden, Albert. *Catalogue of Early Telescopes*. Giunti, Florence: 1999.

_____. "The Astronomical Telescope." *Nuncius* 1 (1976): 13–36.

_____. "The Telescope in the Seventeenth Century." *Isis* 65 (1974): 38–57.

Willach, R. "New light on the invention of the achromatic telescope objective." *Notes and Records of the Royal Society of. London* 50 (1996): 195–210.

_____. "The Development of Lens Grinding and Polishing Techniques in the First Half of the 17th Century." *Bulletin of the Scientific Instrument Society* 68 (2001): 10–15.

_____. "The Development of Telescope Optics in the Middle of the Seventeenth Century." *Annals of Science* 58 (2001): 381–398.

_____. "James Short and the development of the reflecting telescope." *Journal of the Antique Telescope Society* 20 (2001): 3–18.

Zirker, J.B. *An Acre of Glass: A History and Forecast of the Telescope*. Baltimore: Johns Hopkins University Press, 2005.

Dedicated Journal

Journal of the Antique Telescope Society

Some Useful On-line Resources

Abrahams, Peter. "The history of the telescope and the binocular." [Updated 31 August 2008; cited 11 December 2008]. Available from http://www.europa.com/~telscope/binotele.htm. An extensive Web site cataloguing most of the known literature on the history of the telescope.

Bolt, Marvin, Eugene Rudd, and Duane Jaecks. "Dioptrice: Refracting Telescopes prior to 1775." [Cited 11 December 2008]. Available from http://historydb.adlerplanetarium.org/dioptrice/. Ongoing research about telescopes made prior to 1750.